MW00464572

Janus Earth Science

GLOBE FEARON
Pearson Learning Group

CONTRIBUTORS

Susan Echaore-Yoon
Susan Kaschner Jagoda
Winifred Ho Roderman
Mary K. Friedland
Antonio Padial

CONSULTANTS

Alan Gould, Lawrence Hall of Science, University of California, Berkeley
Gilbert Yee, Fremont, California
Doris Sloan, Ph.D., University of California, Berkeley
E. Jan Null, National Weather Service, Fremont, California

REVIEWER

Peggy A. Prazma, Piedmont High School, San Jose, California

ARTISTS

Margaret Sanfilippo, Ellen Beier, Rawn McCloud, Nancy Kirk

PHOTO CREDITS

cover: NASA, 3: NASA, 49: U.S. Geological Survey, 73: Doug Falk, 73: R. Collins Bradley, Santa Fe Railway, 95: Sam S. Pierson, Photo Researchers, 110: David Petty, Photo Researchers, 134: Grant Heilman, 134: Grant Heilman, 134: Grant Heilman, 134: Grant Heilman, 137: Lawrence Migdale, Stock Boston, 154: Ulrike Welsch, Photo Researchers, 141: Grant Heilman, 161: Grant Heilman, 168: The Field Museum, Chicago, IL Neg. # Geo 85637 Photo by John Weinstein

ISBN 0-835-91386-4
Printed in the United States of America
8 9 10 11 12 13 06 05 04 03 02

1-800-321-3106
www.pearsonlearning.com

Contents

Introduction

In this book, you will learn about some of the things that scientists study. You will learn some of the things that scientists have discovered. You will explore and discover facts the same way scientists do, by experimenting, observing, and recording. And you will learn scientific information about the world that you can use right away.

The World Around Us

Earth science is the study of the planet on which we live. Everything that happens on, around, and inside Earth is important to your life.

What causes changes to Earth's surface? How are mountains formed? What are some of Earth's resources? Why is the weather different in your part of the world from the weather in other places?

Earth scientists study our planet. They have discovered some of the answers to those questions. Because of what scientists have discovered, we know important facts about our planet and about other planets in our solar system. And we know about the land we live on, about the weather around us, and about the resources we need to stay alive.

Scientists can tell us how the land around us is always changing. They can tell us about other planets in our solar system and why the seasons change. They can tell us what the weather will be like tomorrow and where rain comes from. And they can tell us where to find Earth's resources and how we can keep them from running out.

In this book, you'll learn about the planet on which you live. As you study Earth, you'll use the same methods that scientists use: You will observe the world around you and record what you learn. And you will discover many of the things that scientists can tell us about the world we live in.

THE SOLAR SYSTEM

What is in our solar system? What are stars? What are planets? How is Earth different from other planets? How is the sun important to our life? What have scientists learned from space travel? In this section, you'll learn many facts about our solar system. And you'll learn how important the other things in our solar system are to our life here on Earth.

Contents

Introduction

Picture this:

It's night. You're walking down a road with a friend. Suddenly your friend says, "Hey! Look at the sky!"

You look up and what do you see?

Stars! Black space!

Above you are millions of stars, shining in black space.

Did you ever wonder about those stars?

Some of those stars are very far away. Some are closer to us. The sun is the closest star. Some of the things that look like stars are not stars. They are really worlds, something like the world we live on.

Those worlds, our world, and the sun make up our solar system. What is the solar system? Is there life on any of those worlds? What is the star that's in our solar system?

In this section, you'll learn the answers to those questions. You'll find out other things about the solar system. And you'll see photographs that were taken in space.

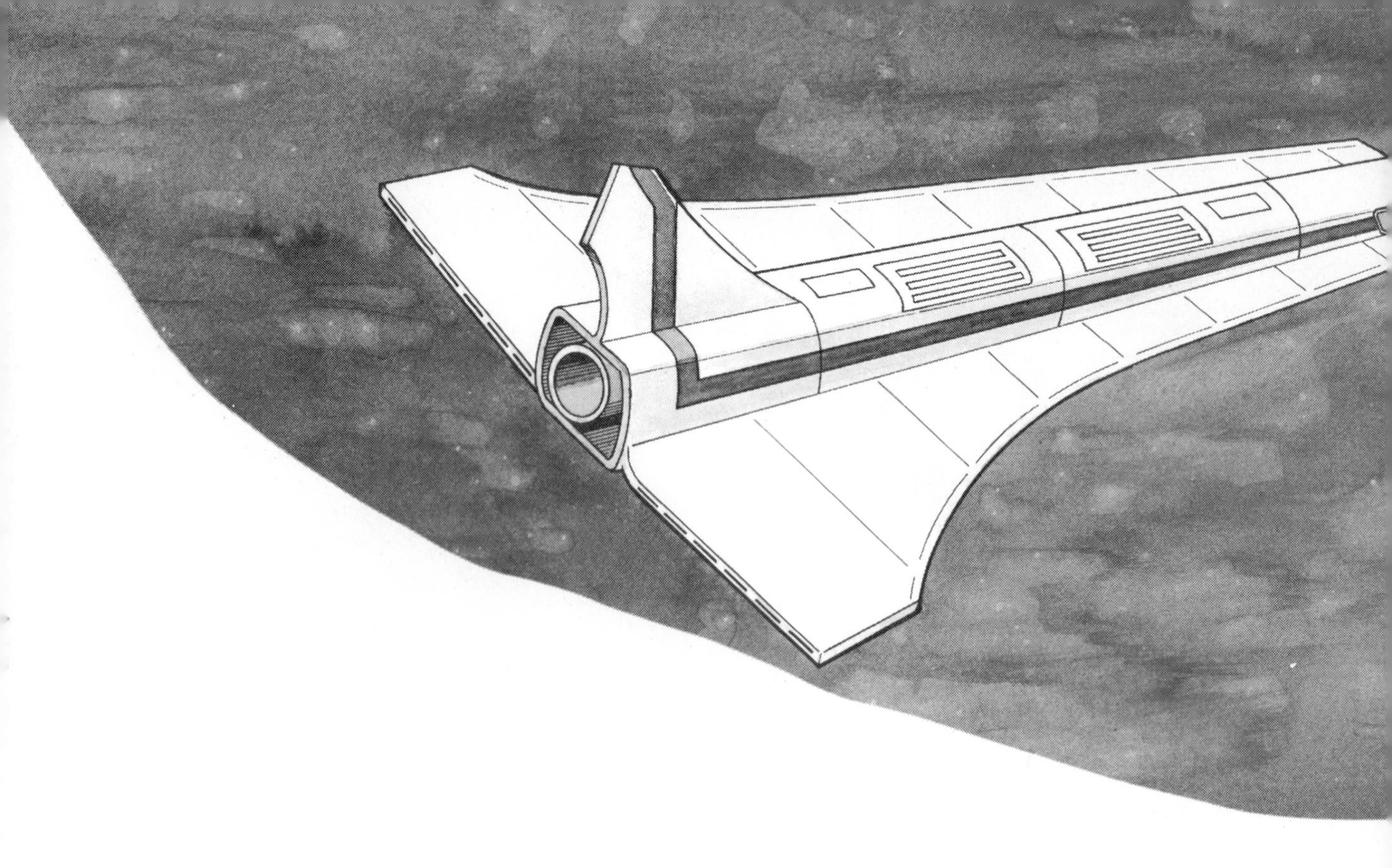

Unit 1

What's Your Space Address?

Suppose someone asks, "Where do you live?"

You would probably say your house number, the name of your street, and the name of your town or city.

But you also have a space address. You live on a **planet**. That planet is in a star system. And that star system is in a certain group of stars.

- What planet do you live on?
- What star system is your planet in?
- What group of stars is your star system in?

You'll learn the answers in this unit.

Before You Start

You'll be using the science words below. Find out what they mean. Look them up in the Glossary that's at the back of this book. On a separate piece of paper, write what the words mean.

1. **galaxy**
2. **star**
3. **star system**

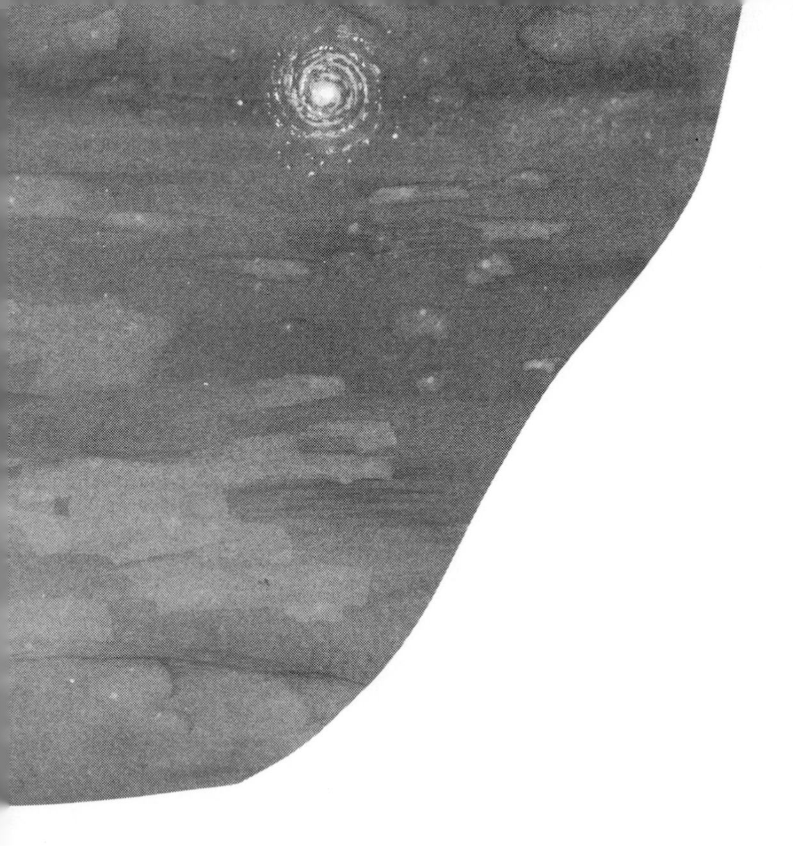

Coming Home

Imagine this:

You are traveling home to Earth from outer space.

You're in a spaceship. You're far away from the huge group of stars that you live in. The space around you is almost completely black. Why do you think the space is almost completely black?

Right! The nearest star, or sun, is far away. Very little of its light reaches you.

Your ship is heading toward a dim light far in the distance. That light is shaped something like a circle.

As you get closer to the light, you see that it's made up of billions and billions of stars. Those stars make up your home galaxy—the group of stars you live in.

Do you know the name of your galaxy? What is it?

Galaxy Ahead!

The galaxy we live in is the *Milky Way.* Look at the picture above. It shows what the Milky Way might look like from your spaceship. What does it look like to you?

From space, the Milky Way galaxy looks like a giant whirlpool of stars.

But from Earth, the Milky Way looks like a huge, white path of stars. That's because we can see only part of the Milky Way from Earth. We can see it only at night.

Why do you think people gave this path the name *Milky Way*?

Look at the sky tonight. See if you can find the path that we call the Milky Way. You'll see part of the galaxy you live in.

Star Ahead!

Imagine that your spaceship is now in the Milky Way. You are looking for a certain star toward the edge of the Milky Way. What star do you think you're looking for?

You're looking for our sun. Our sun is a star. It is just one of billions of stars in the Milky Way.

Stars are huge balls of superhot, burning gases. Stars give off light and heat that can be seen and felt millions of miles away.

Now when you look at the sky, you'll know this: Some of the things that look like stars are *not* stars. What do you think they are?

Right! They are worlds—or *planets.*

Star System Ahead!

Your spaceship gets closer to our sun. You see this: nine planets close to a huge, superhot, burning ball. You're seeing our star system.

The main parts of our star system are the sun and the planets that stay close to it. What's the name of our star system?

Right! Our star system is called the **solar system**.

There are many star systems in the Milky Way. Some may have more planets than ours. Some may not have as many. Some may even have planets like Earth where things can live.

Do you think there's life like ours in another star system? Why or why not?

Home!

Your space trip is ending! You're heading toward a blue and white planet. A small white moon hangs above that planet.

You are looking at Earth, the third planet from the sun. You are *home*.

On a separate piece of paper, write your space address. First, write the name of your planet. Next, write the name of the star system your planet is in. Then write the name of the galaxy your planet is in.

Review

Show what you learned in this unit. Finish the sentences. The words you'll need are listed below. One word will be left over.

galaxy	Milky Way	Earth
star	solar	planet

1. A huge group of stars is called a ________.
2. Our sun is a ________.
3. Our sun and nine planets are part of the ________ system.
4. The planet ________ is part of the solar system.
5. Our galaxy is called the ________ ________.

Check These Out

1. Make a Science Notebook for this section. Use it to keep a record of what you learn about the solar system. Keep your list of glossary words and their meanings in the notebook. You can put anything else you learn about the solar system in your Science Notebook too.
2. Find out about stars. How are they born? How can we tell how old they are? How do they die? Tell your class what you find out.
3. Do this one night: Pick a group of stars that's above a tree or a building. Draw a picture of the group of stars and the tree or building. A month later, stand in the same spot. Look for that group of stars and the tree or building. Draw another picture of the stars and the tree or building. Is the group of stars in the same place, or has it moved?
4. Make a poster of the sun, Earth, and moon. Show how they look together in space. Color your poster.
5. As you work through this section, you may want to find out more about the solar system. You can find out more by looking in an encyclopedia or by getting books about the solar system from a library. You can also talk to an expert, such as an astronomer or an astronaut.

   Here are some things you may want to find out:
   - What is the universe? What is a constellation?
   - What is a black hole?
   - What are quasars and pulsars?
   - What is astronomy? What is astrology?

Unit 2

The Planet We Live On

Look around you. What do you see? People? Trees? Birds? All of those things are alive. And they live on the planet Earth.

Earth is just the right distance from the sun. It's not too hot or too cold. That's why people and other things can live on Earth. Earth is the only planet we know of that has life.

- What is our planet like?
- Why don't things on Earth fall into space?
- Why does Earth have night and day?

You'll learn the answers in this unit.

Before You Start

You'll be using the science words below. Find out what they mean. Look them up in the Glossary. On a separate piece of paper, write what the words mean.

1. **axis**
2. **gravity**
3. **rotate**

NASA photograph

Planet Earth in space

Planet Earth

This photograph shows what Earth looks like from a spaceship. The photograph was taken by a **NASA** spaceship as it traveled to the moon. (NASA is our country's program to explore space.)

What does Earth look like?

If you were on that NASA spaceship, you'd see a huge blue and white ball. The blue parts are the oceans that are on Earth. What do you think the white parts are?

Right! The white parts are clouds. The clouds are part of Earth's **atmosphere**. (Earth's atmosphere is the air that covers Earth.)

Earth is not perfectly round. The parts we call the "top" and the "bottom" are a little flat. We call the top part the North Pole. What do you think the bottom part is called?

Right! It's called the South Pole.

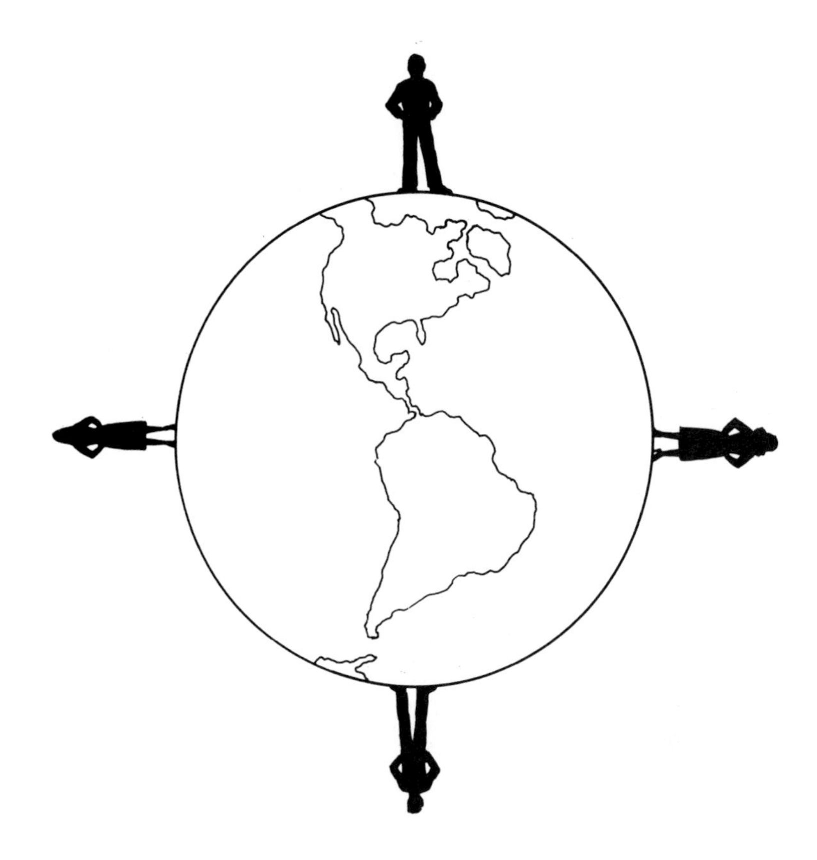

Sticking to Earth

Toss a ball in the air. What happens to it?

Jump up. What happens to you?

Both you and the ball land on the ground. Now toss a light thing, such as a feather, in the air. What happens to that thing?

No matter how heavy or light something is, it will always fall to the ground. What makes things fall down instead of staying up in the air?

Right! Gravity.

Gravity is the strong pull that holds everything and everyone to Earth. For example, gravity holds our atmosphere close to Earth. Gravity keeps that atmosphere from flying off into space.

What do you think would happen to you if Earth didn't have gravity?

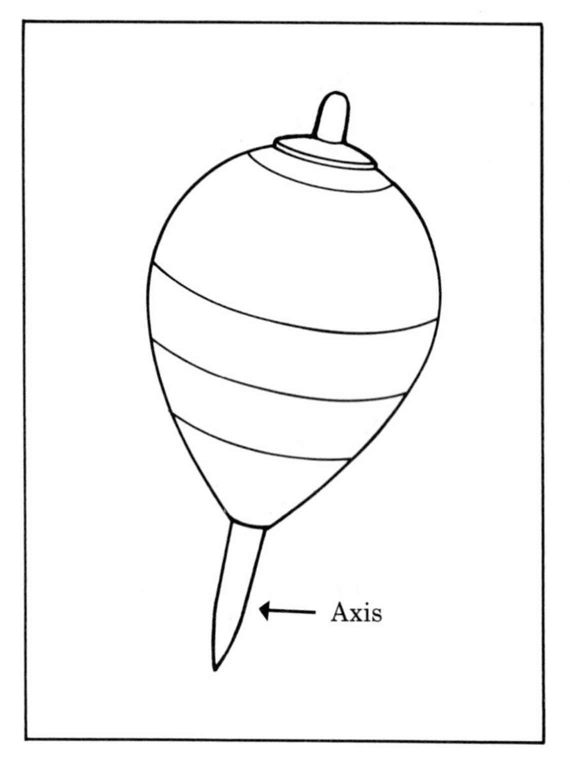

Top

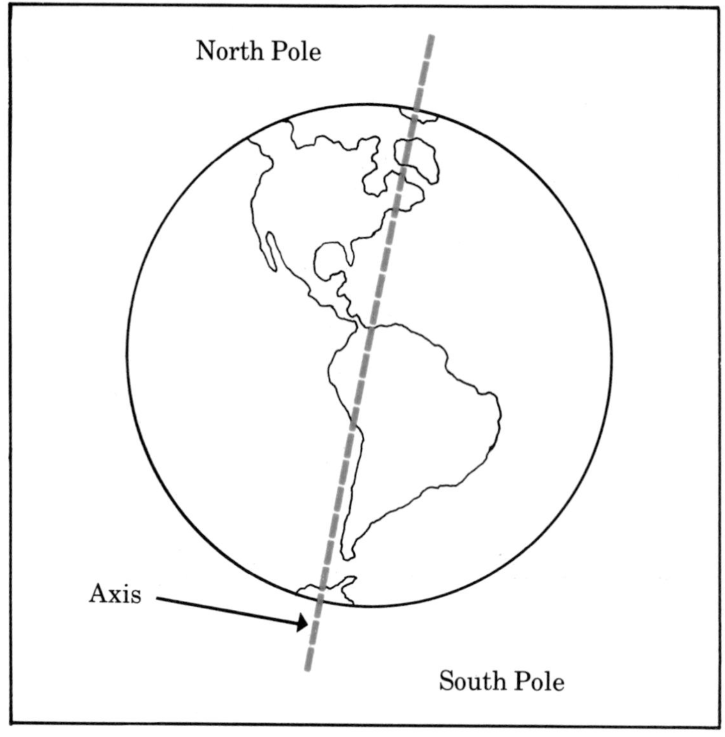

Earth

Earth Spins

Right now, our planet is moving. Do you feel it moving? Do you see it moving? Probably not. But Earth *is* moving.

How do you think Earth is moving?

One way that Earth moves is to *rotate*. In other words, it spins.

How does Earth spin? Imagine that there's a line through the center of Earth. The line goes from the North Pole to the South Pole. That imaginary line is called an *axis*. Earth spins around that axis.

You can see how Earth rotates by spinning a top like the one in the picture. The top has a stick in the middle. When you spin the top, it rotates around the stick. Earth rotates around its axis like that.

Now look at the pictures of the top and of Earth. Is the stick in the top pointing straight up? No. The stick **tilts**.

Is the axis of Earth pointing straight up or does it tilt?

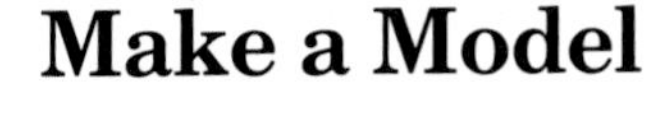

Make a Model

Make a model of Earth. Then make the model rotate around its axis. You will need these materials:

- One sharp pencil
- One Styrofoam ball

1 Carefully push the pencil through the center of the ball. Push the pencil so that it comes out at the opposite end of the ball. The ball is like Earth. What is the pencil like?

Right! The pencil is like Earth's axis.

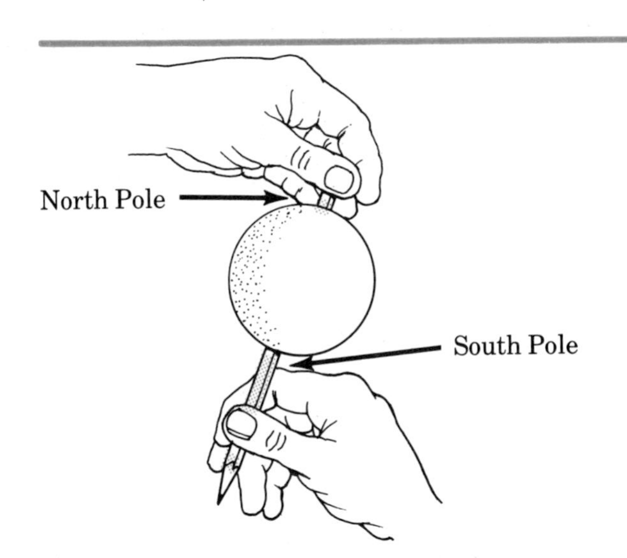

2 Hold the pencil so the eraser is pointing up. What is the top of your model Earth like?

Right! It's like the North Pole. And the bottom is like the South Pole.

Tilt the pencil the way the one in the picture is tilted. Why would you tilt the pencil?

Right! You tilt the pencil because Earth's axis tilts.

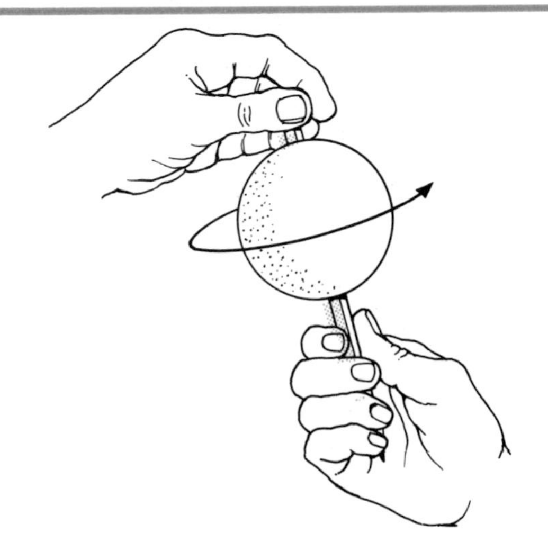

3 Hold the top of the pencil loosely with one hand. Then, with your other hand, slowly turn the pencil to the right. What happens to your model Earth?

Right! It rotates. Your model rotates just as Earth rotates on its axis.

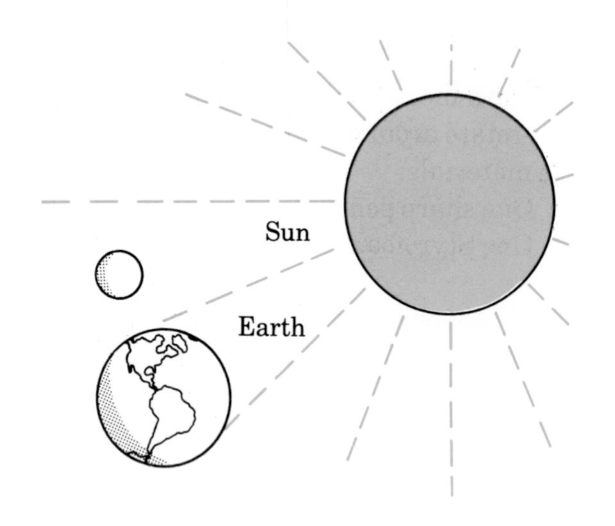

Night and Day

The picture at the top shows Earth and the sun. One side of Earth faces the sun. The sun shines on that side. It is day on that side.

Look at the side that faces away from the sun. The sun can't shine on that side. Is it day or night on that side?

You can see how night turns into day. Use these materials:

- Your model of Earth
- One pin
- One lamp

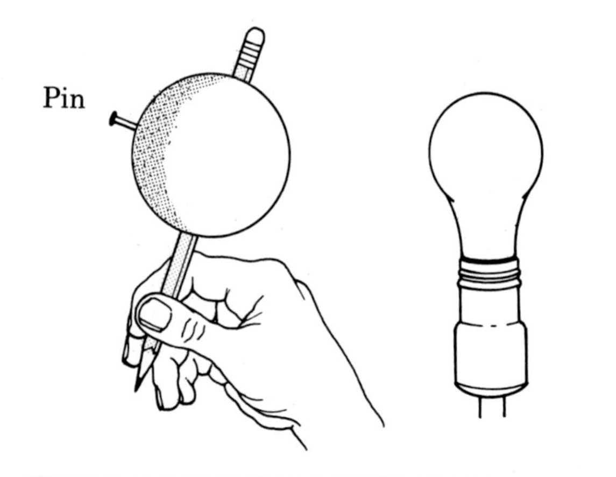

1 Put a pin in the model. Then turn on the lamp. The lamp will be the sun.

Hold your model of Earth the way the model in the picture is held. The side with the pin faces away from the lamp (your sun).

Is it day or night on that side?

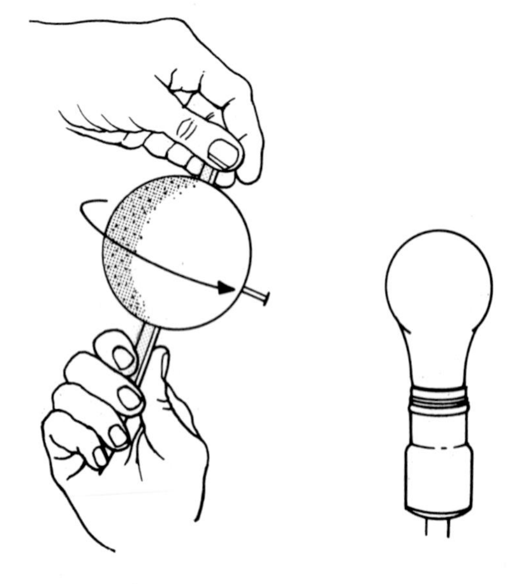

2 Slowly rotate the model to the right. Stop when the pin faces the lamp. Is it day or night on that side?

Now rotate the model until the pin faces away from the lamp. Is it day or night on that side?

Your model rotated completely. The pin went from night to day to night again.

How many hours do you think it takes Earth to rotate completely?

Right! It takes Earth 24 hours to rotate completely.

Review

Show what you learned in this unit. Match the words in the list below with the correct clues.

axis	Earth	atmosphere
rotate	gravity	South Pole
day	North Pole	night

1. The planet we live on
2. Air that covers Earth
3. When your part of Earth faces away from the sun
4. When your part of Earth faces the sun
5. The "top" of Earth
6. To spin
7. The strong pull from the center of Earth
8. The imaginary line through the center of Earth
9. The "bottom" of Earth

Check These Out

1. There are many stories about how Earth began. Some are old folktales. Tell a story about how Earth began. You can write the story or you can act it out for the class.
2. Do you know you can have a birthday two days in a row? Find out about the international date line. Then you'll see how that can happen.
3. You learned that Earth is round. But some people believe Earth is flat. How would you explain to them that Earth *is* round?
4. Here are more things you may want to find out:
   - Who was Isaac Newton? What ideas did he give to science?
   - What is the summer solstice? What is the winter solstice?
   - What is the spring equinox? What is the autumn equinox?
   - When it's 6 a.m. in New York, it's 12 noon in Paris, 8 p.m. in Japan, and so on. Why is the time different in different places?

Unit 3

Our Moon

Did you ever go fishing at night when the moon was full? Remember how bright the moon was? It was so bright, you didn't need a flashlight. Where did the light come from?

Right! It came from the moon. The moon is a little like a planet. But the moon shines, and it doesn't always look round.

- Why does the moon shine?
- Why does the moon seem to change its shape?
- How does the moon move?

You'll learn the answers in this unit.

Before You Start

You'll be using the science words below. Find out what they mean. Look them up in the Glossary. On a separate piece of paper, write what the words mean.

1. **reflect**
2. **revolve**
3. **satellite**

Earth's Satellite

The moon is like a small planet. It's made of the same kinds of things that our planet is made of. (Planets are made of things like rock, metals, and gases.)

The moon is Earth's satellite. A satellite always stays close to something that's larger than it is.

Our moon always stays close to Earth. Why do you think it stays close to Earth?

Earth is larger than the moon. Earth's gravity is strong. It pulls on the moon. So Earth's gravity keeps the moon close to Earth.

The moon and Earth are very different.

Earth has oceans full of water. The moon has no water on it.

Earth has an atmosphere around it. The moon has no atmosphere at all.

Imagine that you are on the moon. What would the moon look like? Describe what you think you would see.

Moonshine

You learned that the moon shines. Look at the picture at the top of the page. Where does the moon's light come from?

Yes! The moon's light comes from the sun. The sun lights up the moon. You can make a model showing how that happens. You will need these materials:

- One Styrofoam ball
- One pencil
- One globe of the world
- One lamp (The lamp should be taller than the globe.)

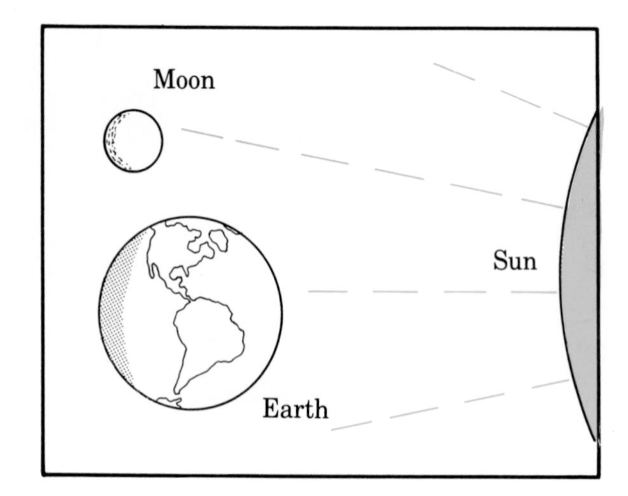

1 Make a model of the moon, using the Styrofoam ball and a pencil.

2 Put the globe about a foot away from the lamp. Turn on the lamp. Turn off all other lights in the room.

3 Hold your model of the moon above the globe. Look at that model. What happens to it?

Yes! One part of your model is lit up. It's lit up by the lamp.

That's what happens to the moon. The sun shines on it and lights it up. We see the moon at night because the sun lights it up.

The moon is like a dull mirror. It reflects the sun's light. And some of that light comes to Earth.

Planets also reflect the sun's light. Suppose you were on the moon. What would Earth look like in the sky?

Right! Earth would shine and reflect the sun's light. Earth would shine like the moon!

NASA photograph

Does the Moon Have Gravity?

Picture this:

You're in a spaceship that just landed on the moon. You're about to step out of that spaceship. Can you walk on the moon? Or will you fly off into space?

Right! You can walk on the moon.

The moon has gravity. Its gravity will pull on you, so you won't fly off into space.

But as you walk on the moon, you'll feel lighter. That's because the moon's gravity is weaker than Earth's.

You learned that the moon has no atmosphere. Why do you think it has no atmosphere?

Right! The moon has no atmosphere because its gravity is weak. Its gravity is too weak to hold an atmosphere around the moon.

Even though the moon's gravity is weak, it pulls on Earth. For example, it pulls on Earth's oceans.

What happens to our oceans when the moon's gravity pulls on them?

Right! The tides go in and out.

The Moon Moves

You learned that the moon is a satellite. And that satellites stay close to a larger thing.

Satellites are always moving. They move around the thing that they stay close to. They *revolve* around it.

What does the moon revolve around?

Right! The moon revolves around Earth. Find out how. You'll need these materials:

- Your model of the moon
- One globe of the world

1 Hold the model to your left and above the globe.

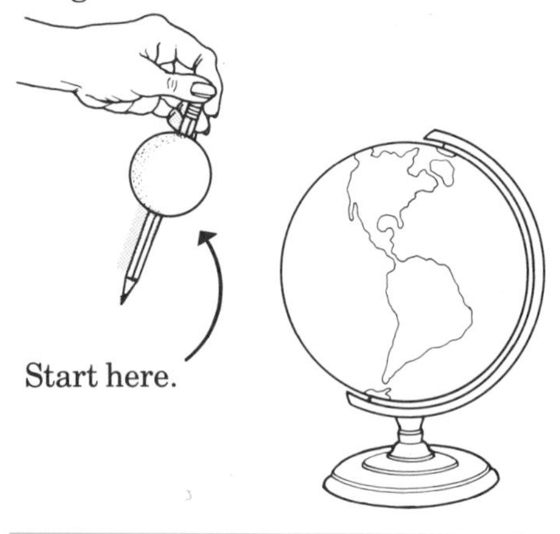

2 Now move the model toward your right.

3 Move the model around the globe in one direction.

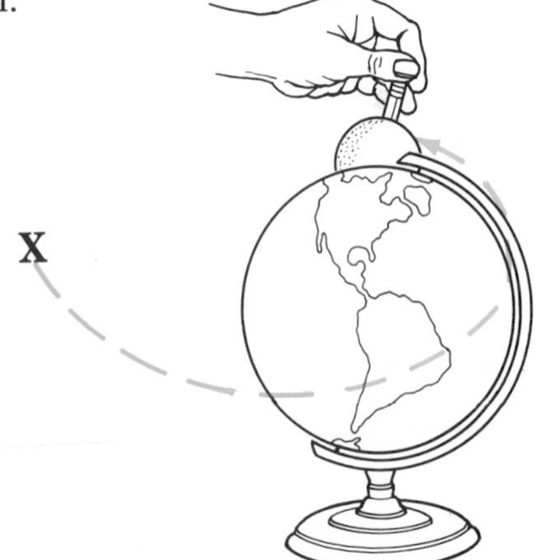

4 Move the model back to where you started.

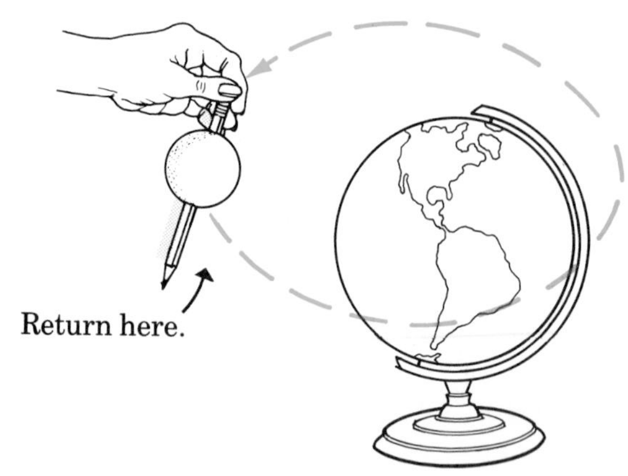

Changing Shapes

The moon is round, but it doesn't always look round to us. Its shape seems to change. Why do you think that happens?

The moon's shape seems to change because the moon revolves around Earth.

Suppose the sun is behind the moon. The side of the moon that faces the sun is lit by the sun. We can't see that side because it faces away from Earth.

The side that faces Earth is in darkness. We can't see that side either. We say the moon then is a *new moon*. (See picture **A**.)

As the moon revolves around Earth, it moves away from the sun. We start seeing the side that's lit by the sun. We see more of it as the moon revolves. Soon we can see half of the side. We then say the moon is a *quarter moon*. (See picture **B**.)

When the moon is halfway around Earth, we can see all of the side that's lit. The moon's shape is now round. We call that moon a *full moon*. (See picture **C**.)

Now the moon starts moving back toward the sun. We see less and less of the side that's lit. Soon we see only half of it. (See picture **D**.) What do we call the moon then?

Right! We call it a *quarter moon*.

When the moon has completely revolved around Earth, the sun is behind the moon again. (See picture **E**.) What is the moon called now?

A
New moon

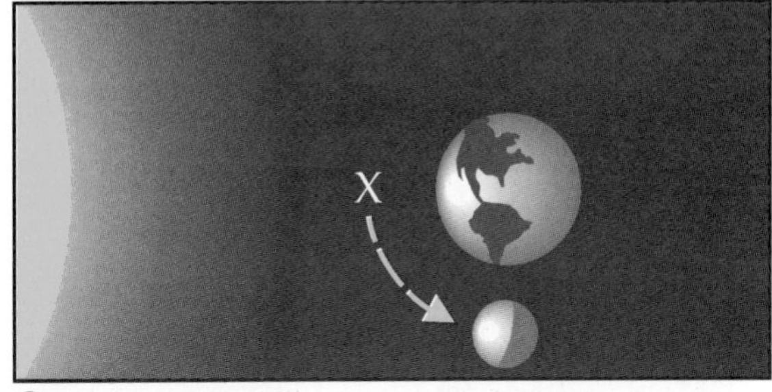

B
Quarter moon

C
Full moon

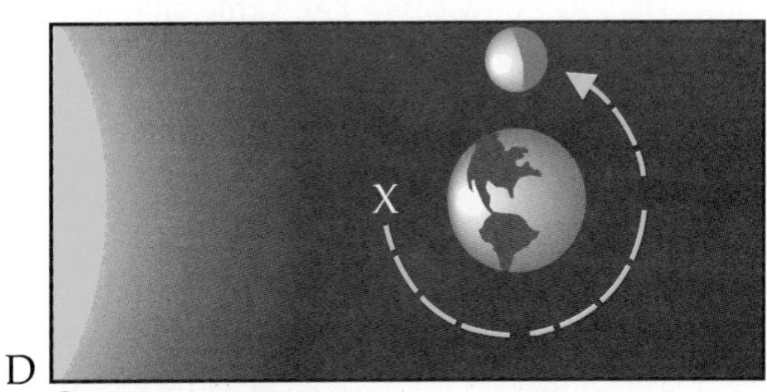

D
Quarter moon

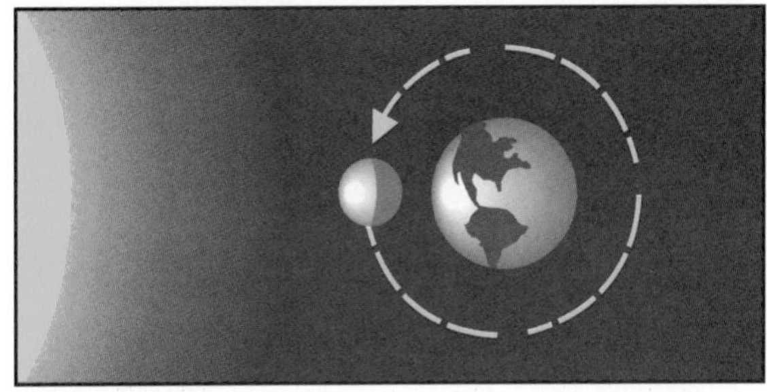
E
New moon

NASA photograph

This side of the moon faces Earth.

Another Way the Moon Moves

Do you know this? The same side of the moon always faces Earth. Why do you think this happens?

Find out why the same side of the moon always faces Earth.

Put a chair in the middle of a room. Stand in front of the chair, facing it.

Now slowly walk around that chair. As you walk, keep the front of your body facing the chair.

What do you have to do to your body so that it always faces the chair?

Right! You have to keep turning your body so that it always faces the chair. You have to rotate your body.

Now think about the moon. What does it do so that the same side always faces Earth?

Right! The moon rotates.

What would happen if the moon didn't rotate?

The Moon Rotates and Revolves

You learned that the moon rotates. And you learned that the moon also revolves. The moon rotates as it revolves around Earth.

Find out how the moon rotates as it revolves. Get these materials:

- Your model of the moon
- One globe of the world
- Pin

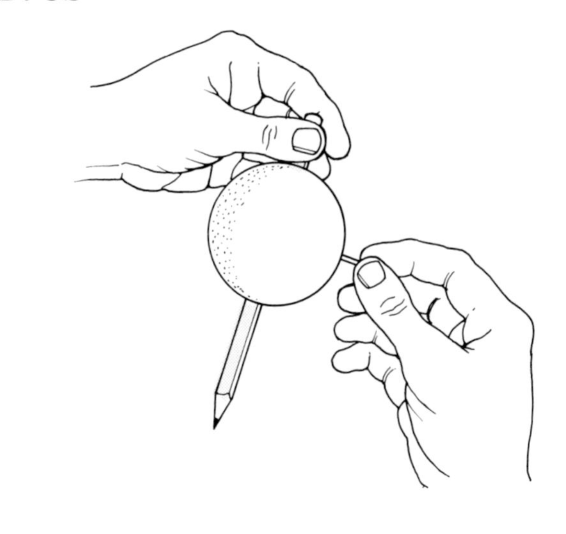

1 Put a pin in the model of the moon. The side with the pin will stand for the side of the moon that always faces Earth.

2 Hold the model to your left above the globe. Then move your model to your right. Revolve your model around the globe.

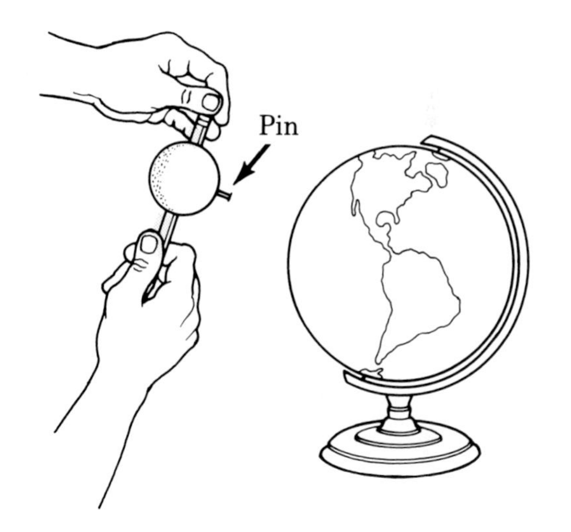

3 As you revolve your model, do this: Very slowly rotate the model so the pin always faces the globe.

That's how the moon rotates as it revolves around Earth.

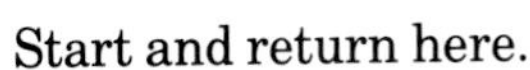

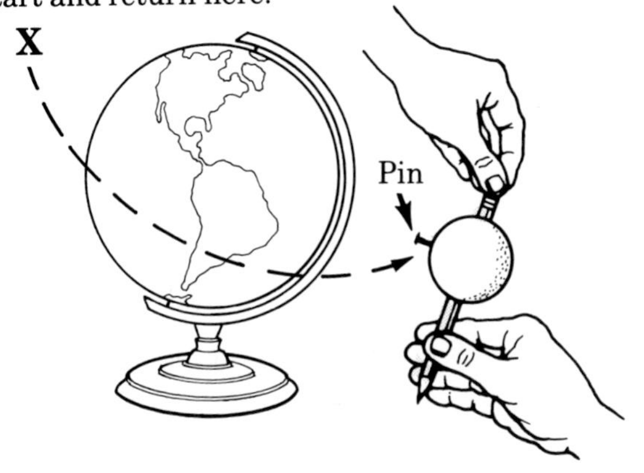

You learned that Earth also rotates. Earth takes 24 hours to rotate completely. (24 hours is a day and a night.)

How long do you think it takes for the moon to rotate completely?

Review

Show what you learned in this unit. Finish the sentences. Match the words on the left with the correct words on the right.

1. The moon is
2. The moon shines because
3. The moon's gravity is
4. The Earth's gravity holds the
5. The moon rotates as it
6. As the moon revolves, its

a. weaker than Earth's gravity.
b. it reflects the sun's light.
c. Earth's satellite.
d. shape seems to change.
e. moon close to Earth.
f. revolves around Earth.

Check These Out

1. How many days does the moon take to rotate completely? How many days does the moon take to revolve around Earth? Find out.
2. Find out about the phases of the moon. Then get a calendar that shows when each phase begins.
   Look at the moon for one month. On a separate piece of paper, draw the moon each time it begins a new phase. Write the date when you see each phase begin.
3. Many songs are about the moon. Find as many as you can. Bring some of those songs to class.
4. Here are more things you may want to find out:
   - What does *lunar* mean?
   - What is a moon, or lunar, eclipse? What causes it?
   - How do scientists think the moon began?

Unit 4

Our Sun

Imagine that it's raining right now. And that it has been raining for days. You think it will never end. You feel low.

Then the sun comes out. Now you feel great!

The sun can make you feel good. But it does more than give good feelings. The sun gives Earth the things we need to live.

- What does the sun give to Earth?
- Why does Earth stay close to the sun?
- How does Earth move in space?

You'll learn the answers in this unit.

Before You Start

You'll be using the science words below. Find out what they mean. Look them up in the Glossary. On a separate piece of paper, write what the words mean.

1. **ellipse**
2. **orbit**
3. **seasons**

Sol

The sun is shaped like a ball. Earth and the moon are also shaped like balls. But the sun is different from Earth and the moon. How is it different?

Right! The sun is a star. Earth and the moon are not stars.

Stars are made up of superhot, burning gases. So the sun is a huge ball of superhot, burning gases. It is the biggest thing in the solar system. And it is in the center of the solar system.

Why do you think the sun is important to Earth?

If we didn't have the sun, nothing would be alive on Earth. The sun gives light and heat that living things need.

But things can't live unless a planet gets the right amounts of light and heat. Earth gets just the right amounts. Why do you think Earth gets the right amounts?

Yes! Earth is just the right distance from the sun.

People have always been interested in the sun. They have called it by many names. One of these names is **Sol**. You've seen these words: *solar system* and *solar energy*. What do you think *solar* means?

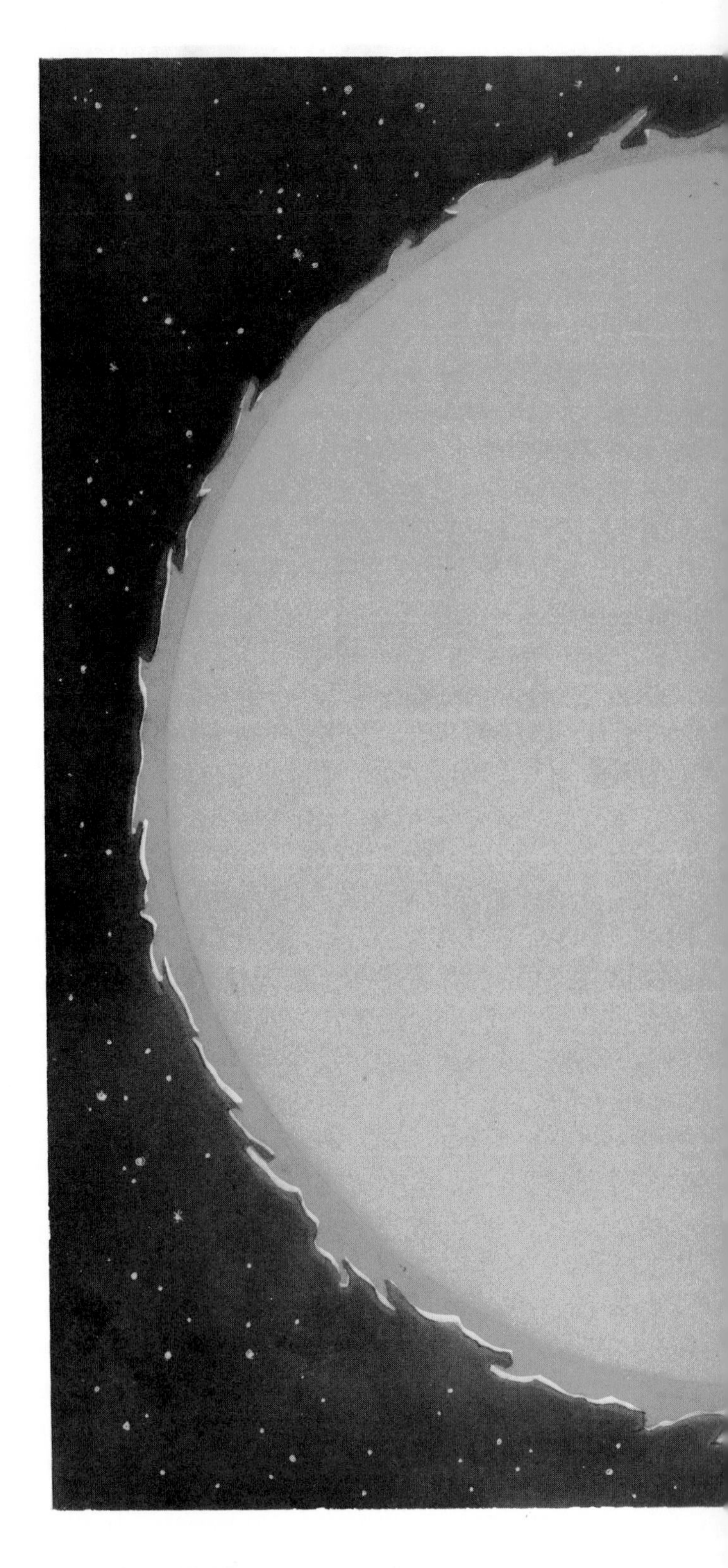

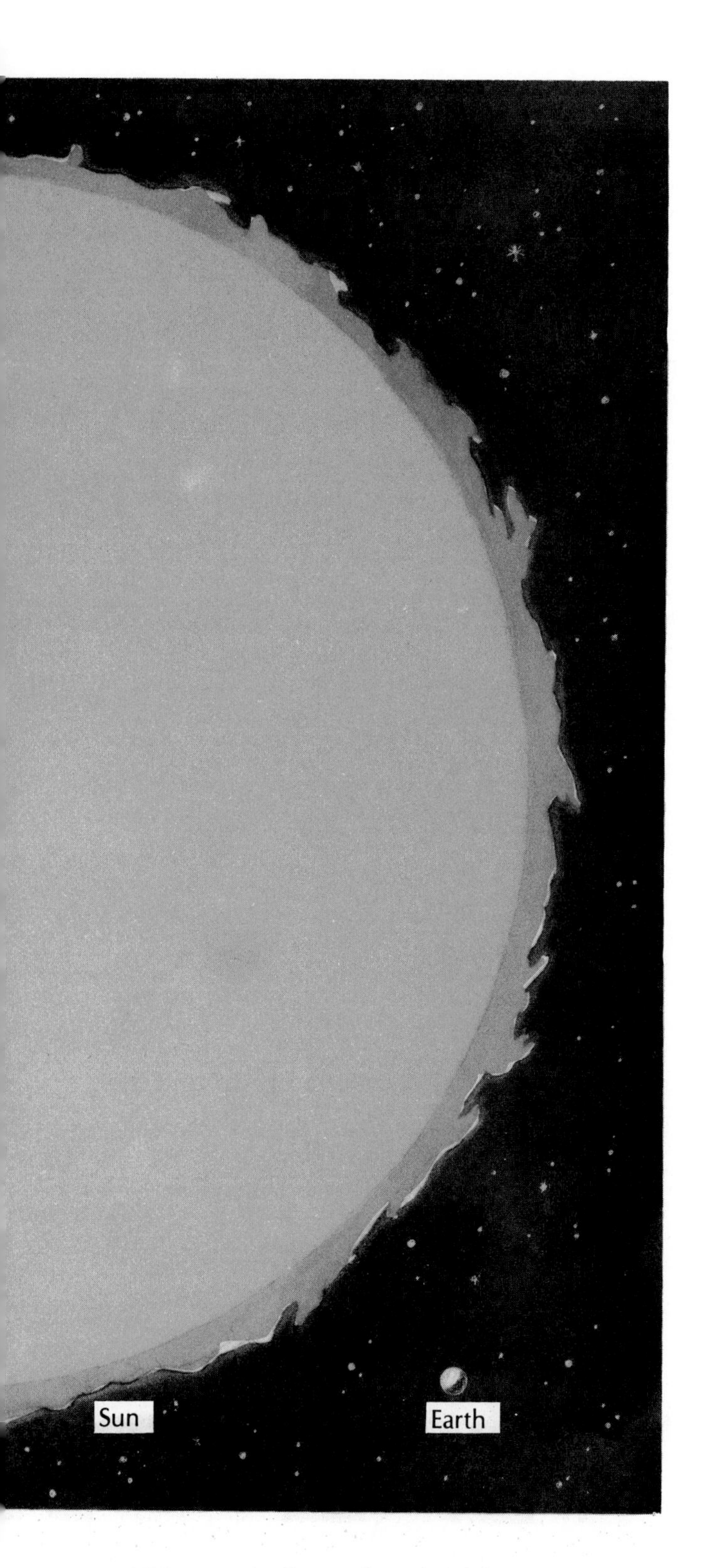

This picture shows the size of the sun.
The small planet next to the sun is Earth.

Our Giant Sun

Suppose you could put all the planets in the solar system together. Which would be bigger: those planets or the sun?

The sun!

The picture on this page shows the sun with Earth next to it. It gives you an idea of how big the sun is.

Here's another way to think about how big the sun is. Get a basketball. Get a piece of bread (or modeling clay). Roll the bread (or clay) into a tiny ball this big: Put that tiny ball next to the basketball.

Look at the sizes of the basketball and the tiny ball. See how big the basketball is next to the tiny ball? That's how big the sun is next to Earth.

Suppose you put a grain of sand next to the tiny ball and the basketball. What do you think that grain of sand shows?

Right! It shows how small the moon is next to Earth and the sun.

Earth and the other planets stay around the sun. Why do you think they do?

Right! The sun has gravity.

It has the strongest gravity in the solar system. So it's the sun's gravity that keeps the planets around the sun.

Moving Around the Sun

You learned that Earth rotates. What other way do you think Earth moves?

Earth *revolves*. It revolves around the sun.

Look at the picture below. The dotted line shows Earth's path around the sun. That path is called an *orbit*.

Look at the shape of Earth's orbit. It's like an oval. That shape is an *ellipse*.

Find out how Earth revolves. You'll need these materials:

- Your model of Earth
- One lamp

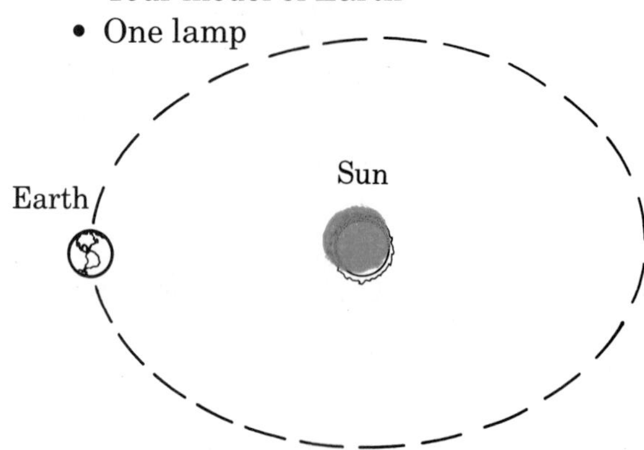

How Earth moves around the sun

1 Set the lamp on a table. The lamp will stand for the sun. Look down at the top of the lamp. Imagine that there's an ellipse around the lamp.

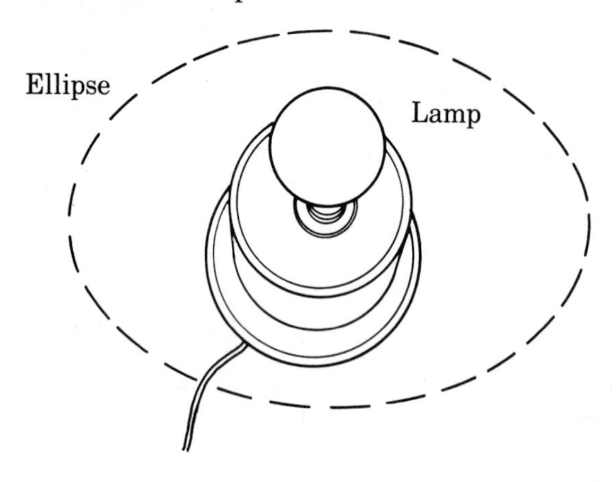

2 Hold your model of Earth the way the model in the picture is held.

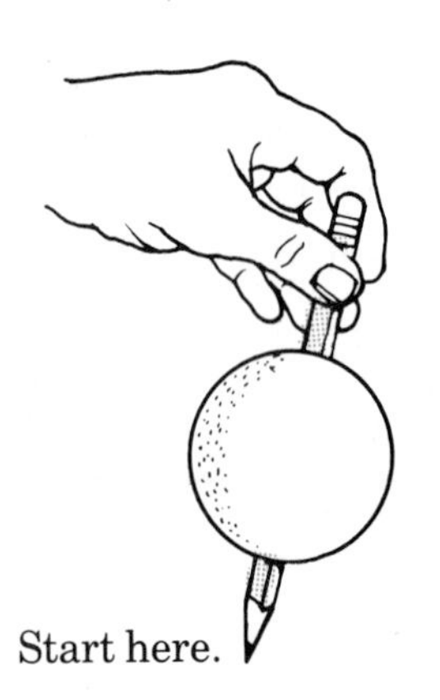

3 Move your model completely around the lamp in an ellipse. That's how Earth revolves around the sun.

How long do you think it takes Earth to revolve around the sun?

Right! It takes Earth one year to revolve around the sun.

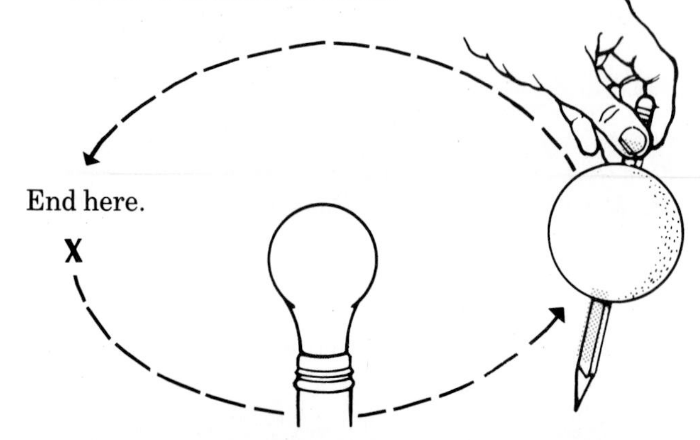

Changing Seasons

Earth has four seasons: summer, fall, winter, and spring. These seasons are always changing.

Summer changes into fall. Fall changes into winter. And winter changes into spring. What does spring change into?

Right! Spring changes into summer.

Why do you think seasons change? Here's why: Earth tilts. And Earth revolves around the sun. As Earth revolves, the seasons change.

Look at picture **A**. It shows Earth when it's summer in the United States.

Now look at picture **B**. Notice where Earth has moved to. It is now fall in the United States.

Look at picture **C**. Earth is halfway around the sun. What season is it now in the United States?

Right! It's winter.

Picture **D** shows Earth as it almost completes its trip around the sun. What season is it now in the United States?

A
Summer

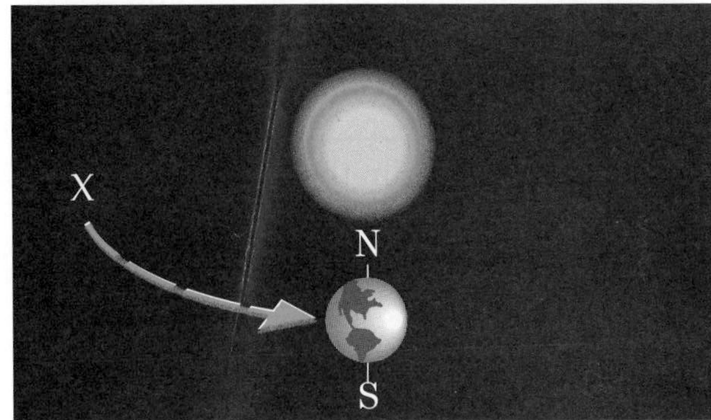

B
Fall

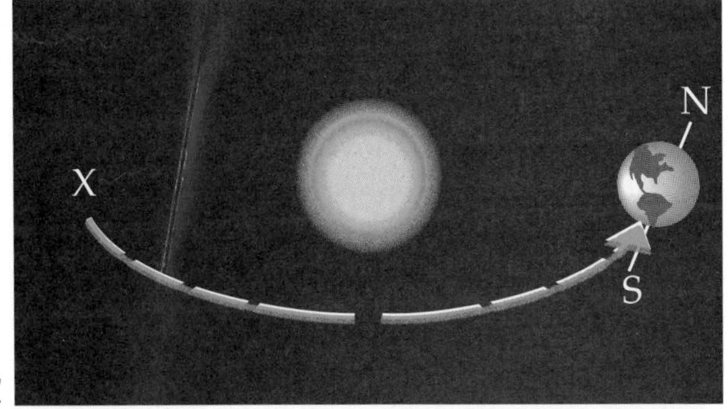

C
Winter

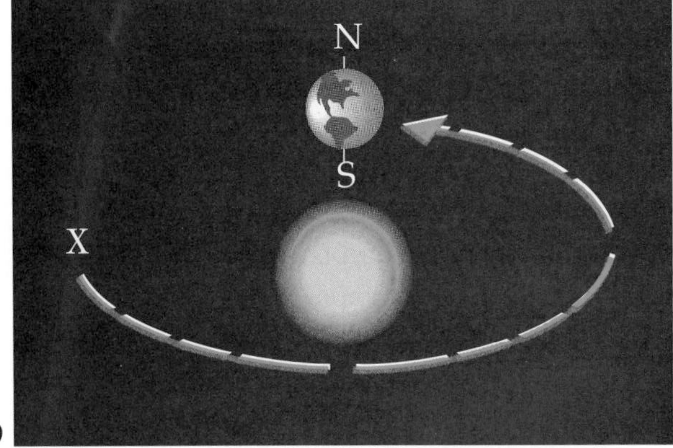

D
Spring

Review

Show what you learned in this unit. Finish the sentences. The words you'll need are listed below.

ellipse	light	seasons
gases	orbit	Sol
gravity	revolves	star

1. The sun is made of very hot burning ________.
2. Our sun is a ________.
3. Earth ________ around the sun.
4. The sun gives Earth ________ and heat.
5. Winter, spring, summer, and fall are ________.
6. The sun's ________ keeps Earth around it.
7. Earth's path is in the shape of an ________.
8. ________ is one name for our sun.
9. Earth revolves around the sun in an ________.

Check These Out

1. When it's summer in the United States, it's winter in Australia. Find out why. Then use the model of Earth and a lamp to show what you learn.
2. What is solar energy? How do people use solar energy? What are new ways that this energy can be used? Find out. Then make a poster showing different ways to use solar energy.
3. Get a partner. Together, find out the answers to these questions:
   - What's the sun made of?
   - How hot is the sun?
   - Why does the sun always stay hot?
   - How was the sun born?
   - Will our sun ever die?

   Now make up a short play about the sun. Pretend the sun is a person. Pretend a TV reporter is asking the sun questions about its life.
4. Here are more things you may want to find out:
   - How far is the sun from Earth?
   - What is a light year?
   - How do scientists think life began on Earth?
   - What is a solar eclipse? How is it different from a lunar eclipse?

Unit 5

Our Solar System

You learned that the sun is the biggest thing in the solar system. You also learned that there are planets, such as Earth, around the sun. But there are many other things around the sun. The sun and all the things around it make up the solar system.

You learned that Earth is always moving. The other things in the solar system are always moving too.

- What are some of the things in our solar system?
- How do these things move?

You'll learn the answers in this unit.

Before You Start

You'll be using the science words below. Find out what they mean. Look them up in the Glossary. On a separate piece of paper, write what the words mean.

1. **collide**
2. **object**

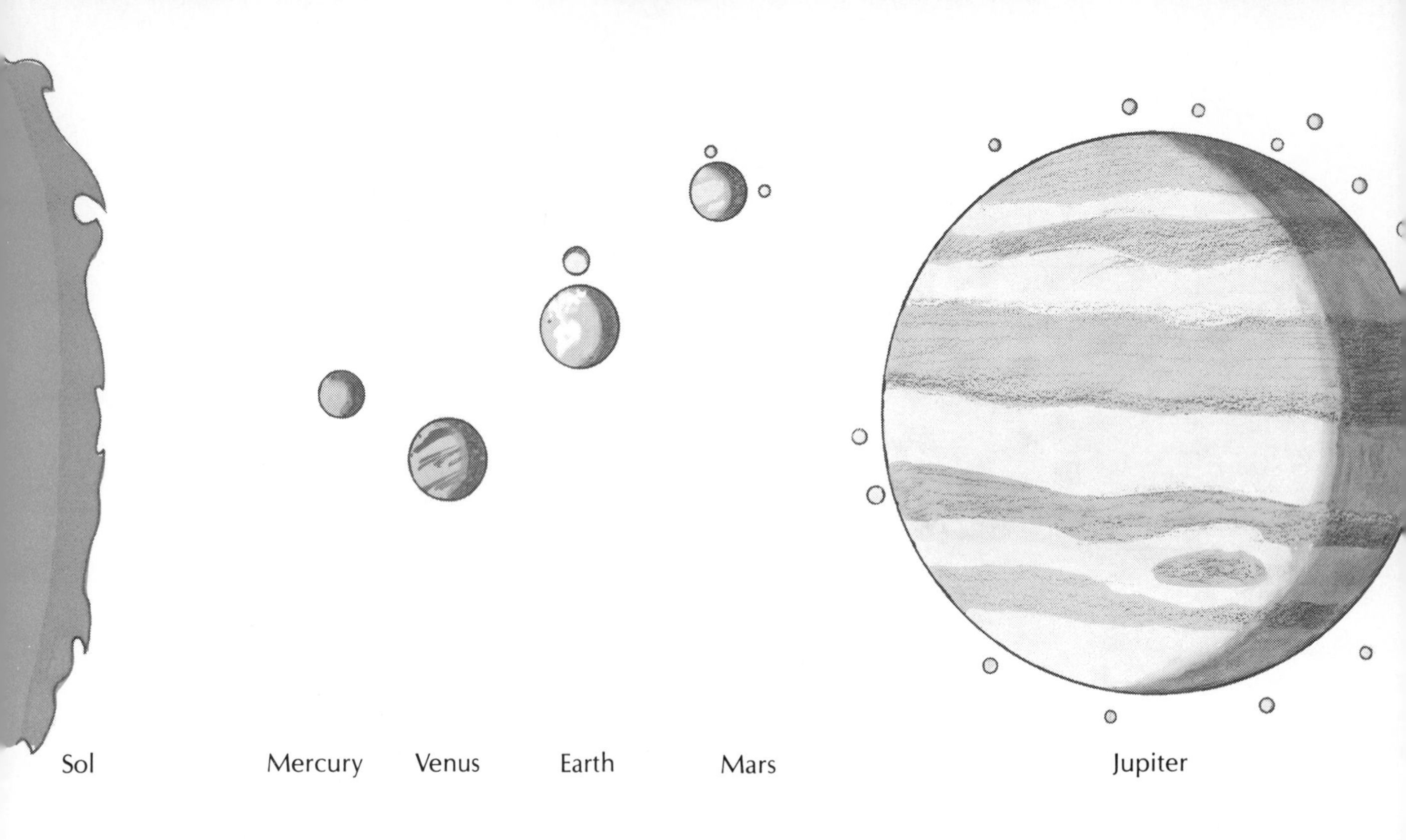

Sol's Planets

Our solar system is made up of the sun and everything that's around the sun. You already know what some of those things are. They are planets.

Scientists know of nine planets in our solar system. The picture above shows those nine planets. It shows the different sizes of the planets. Which is the largest planet?

Right! Jupiter is the largest planet.

The picture also shows the order of the planets from the sun. For example, Pluto is the farthest from the sun. So Pluto is the ninth planet from the sun.

What planet is Jupiter from the sun? (Count the planets, starting with Mercury. Stop when you get to Jupiter.)

Right! Jupiter is the fifth planet from the sun.

Now find Earth. What planet is Earth from the sun?

Right! Earth is the third planet from the sun.

Look at Jupiter again. You'll see many small objects around it. What do you think those objects are?

Yes! Those objects are moons.

Which planets have moons?

Which two planets don't have moons?

Saturn Uranus Neptune Pluto

Going Around in Orbits

You learned that Earth revolves around the sun in an orbit. The other planets also revolve around the sun in orbits.

The time it takes a planet to go around the sun is called that planet's year. How many days are in an Earth year?

Right! There are about 365 days in an Earth year. That's how long it takes Earth to go around the sun.

Other planets have orbits that are longer or shorter than Earth's. Planets with shorter orbits take less time to go around the sun. Planets with longer orbits take more time.

For example, Mars's orbit is longer than Earth's. So it takes Mars longer to go around the sun. A Mars year is almost twice as long as an Earth year.

The planets move at incredible speeds. But they always stay in their orbits. Why don't they move out of their orbits?

If you're not sure, read page 29 again.

The Worlds Turn

All the planets rotate, or spin, around an axis. The time it takes a planet to rotate completely is called that planet's day. How many hours are in an Earth day?

Right! There are 24 hours in an Earth day. That's how long it takes Earth to spin completely around.

Some planets have days that are longer than 24 hours. Some planets have days that are shorter. Here's why some planets have longer or shorter days.

Planets rotate at different speeds. For example, Earth rotates faster than Pluto does. So Earth completes a spin before Pluto does. And an Earth day is shorter than a Pluto day. A Pluto day is 154 hours long. (That's more than six Earth days.)

One Mercury day is as long as 59 Earth days—or about 1,416 hours. One Jupiter day is 10 hours long. Which planet spins fastest—Mercury, Jupiter, or Earth?

JPL/NASA photograph

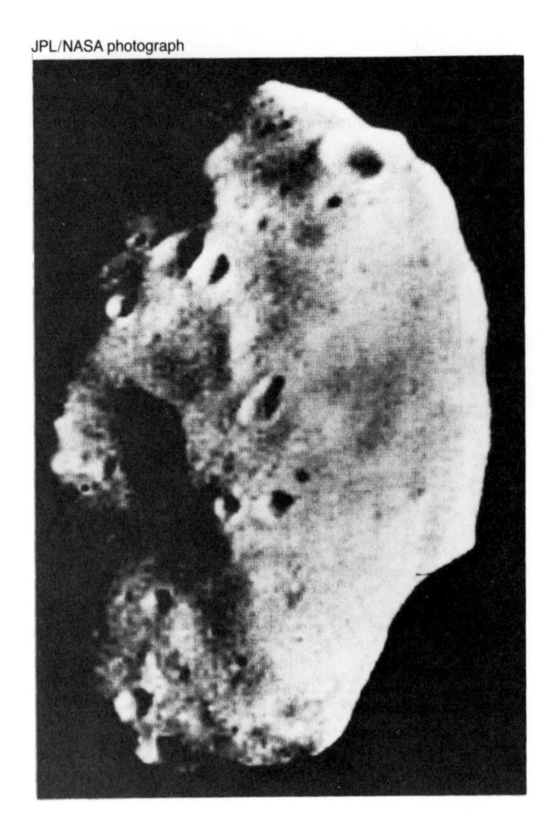

A What an asteroid may look like

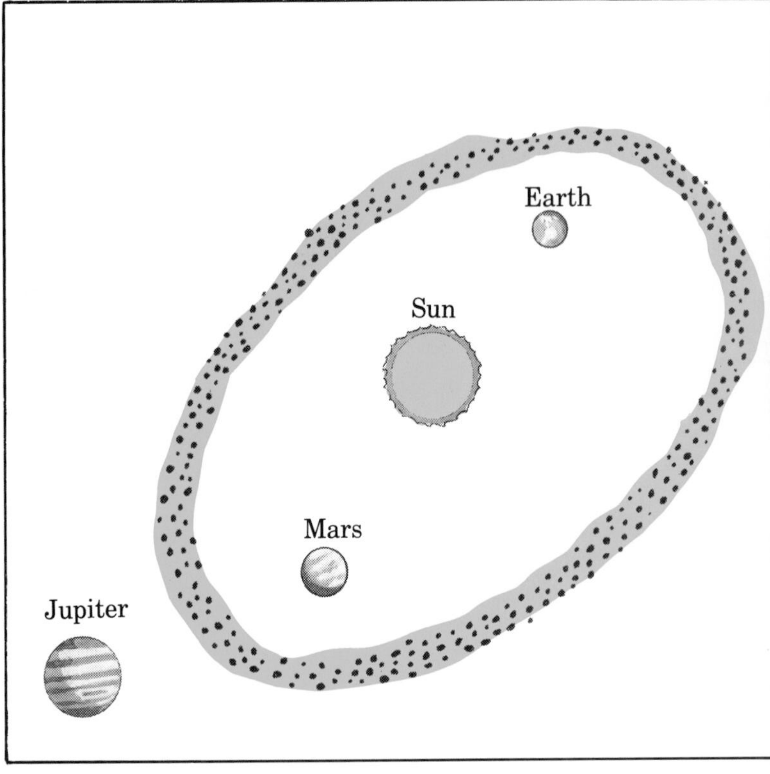

B The asteroid belt

Asteroids

You learned that there are planets around the sun. But there are other smaller objects too. Some of those objects are **asteroids, comets,** and **meteoroids**.

Asteroids are sometimes called "little planets." They are made of some of the same things that planets are made of—rock and metals. Many scientists think that some of the asteroids are pieces of old planets that broke apart.

Some asteroids look like round balls. Some look like jagged rocks. Asteroids can be bigger than a mountain. Or they can be the size of a boulder.

Do you think asteroids move? Or do they stay in the same place in space?

Right! Asteroids move. They revolve around the sun, just as planets do. Each asteroid has its own orbit around the sun.

There are thousands of asteroids in our solar system. Most of them seem to be in a huge, wide ring that goes around the sun. That ring is called the *asteroid belt*. That ring is between Mars and Jupiter.

Look at picture **B**. It shows what scientists think the asteroid belt looks like. Why do you think it's called a belt?

NASA photograph

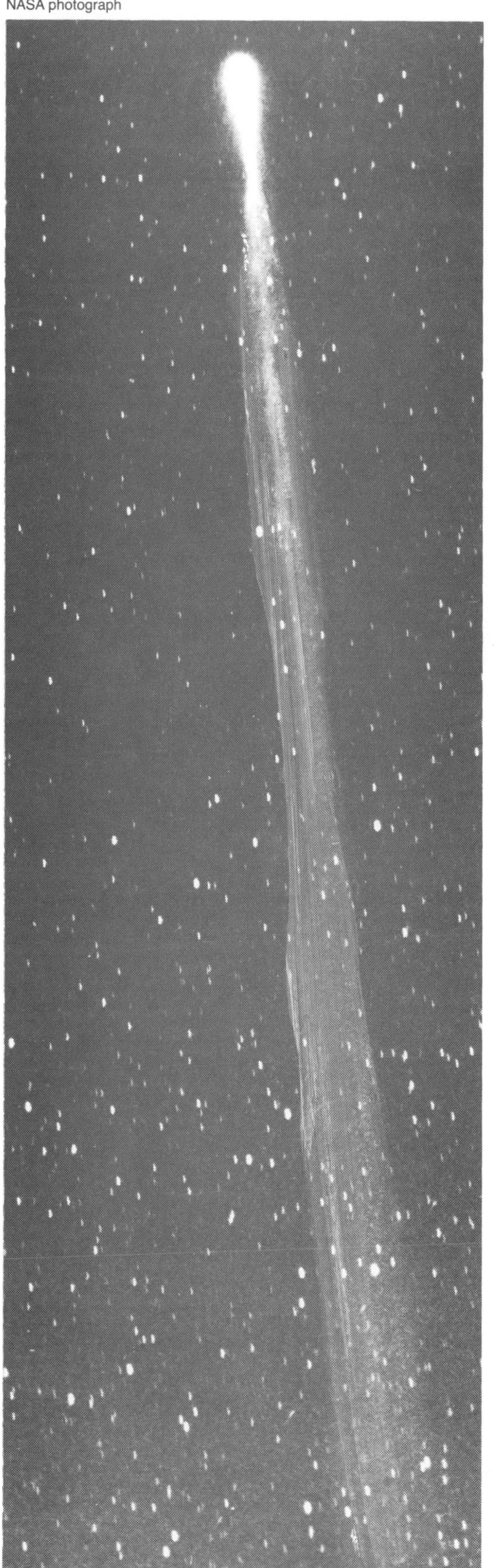

Comet Kohoutek

Comets

The photograph on this page shows a comet. Scientists named that comet *Kohoutek.*

How do you think a comet is like an asteroid?

A comet is like an asteroid in these ways: It is smaller than a planet. It has an orbit.

Comets are balls of ice, gases, dust, and rocks that revolve around the sun.

As a comet moves toward the sun, the heat of the sun melts some of the comet's ice. The melted ice becomes gases that stream away from the moving comet and form a long tail. When that happens, the comet looks something like Kohoutek, the comet in the picture.

Comets have very long orbits. Some comets take thousands of Earth years to complete their orbits. Those comets have orbits that are much longer than Pluto's. (Pluto is the farthest planet from the sun. Pluto takes 248 Earth years to complete its orbit.)

Some comets take less time to complete their orbits. For example, Halley's Comet takes a little more than 75 Earth years to complete its orbit.

As Halley's Comet travels toward the sun, it passes Earth. When that happens, we can see that comet from Earth. Halley's Comet passed Earth in early 1986. When does it pass Earth again? (Hint: Add 75 to 1986.)

NASA photograph

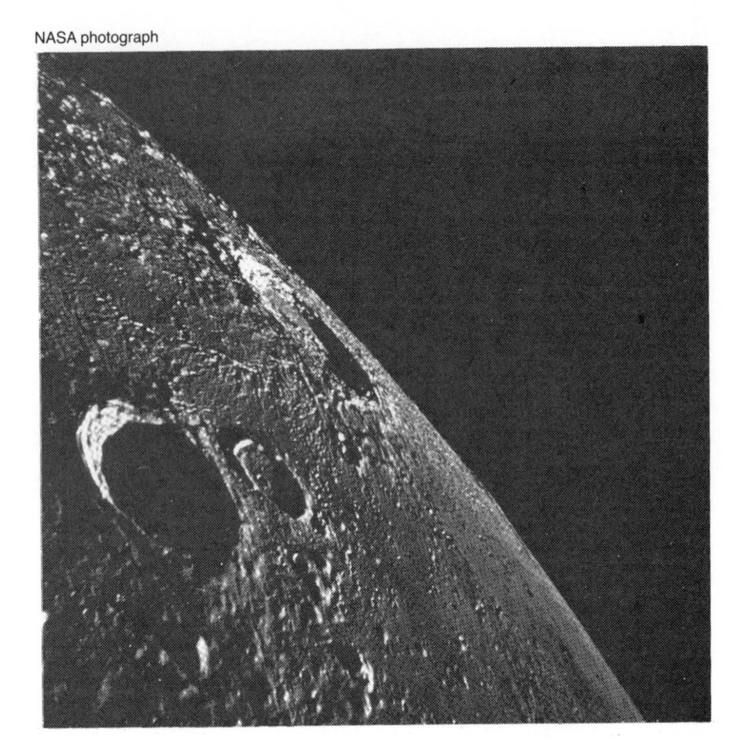

What made these craters on the moon?

Meteoroids

Has this ever happened to you? You're looking at the sky. Suddenly you see a star fall. What do you think that falling star is?

That falling star is *not* a star. It is a meteoroid that is burning up in our atmosphere.

Meteoroids are pieces of rocky material in space. Meteoroids can be as big as boulders or as tiny as specks of dust. There are millions of meteoroids in our solar system.

Meteoroids travel through space at very high speeds. Sometimes they collide with planets or moons. When that happens, they are called **meteorites**.

Look at the picture of our moon. It has often been hit by meteorites. How would you describe the moon?

The moon has many **craters**. The craters were made by meteorites crashing into the moon.

Meteoroids sometimes get close to Earth. But very few ever collide with Earth. That's because of Earth's atmosphere. Meteoroids travel so fast that they burn up in our atmosphere. (When that happens, they are called **meteors**.)

The moon has no atmosphere. So meteoroids don't burn up when they get close to the moon. They crash into the moon instead.

Suppose Earth had no atmosphere. What would happen when meteoroids get close to Earth?

Review

Show what you learned in this unit. Finish the sentences. Match the words on the left with the correct words on the right.

1. The solar system is the sun
2. Planets, asteroids, comets, and meteoroids
3. All planets rotate
4. Comets are balls
5. Meteoroids usually burn up in
6. Asteroids are sometimes called

a. revolve around the sun.
b. at different speeds.
c. and all the objects around it.
d. Earth's atmosphere.
e. of ice, gases, dust, and rocks.
f. "little planets."

Check These Out

1. With your class, make a model of the solar system. Cut out paper models of the sun and each of the nine planets. (Use the picture on pages 34 and 35 to figure out how big each model should be.) Hang the paper models from the ceiling with string.
2. Make a poster showing how far Earth and the other planets are from the sun. Find out how many miles Earth is from the sun. Write the number of miles on the poster.
3. Find out what a meteor shower is. And find out when you might see those showers during the year.
4. Here are more things you may want to find out:
   - How did the planets get their names?
   - How do scientists think the solar system began?
   - Who was Edmund Halley? Tell about his life.

NASA photograph

Unit 6

Exploring Our Solar System

People have always wondered about the stars. At one time, people just explained the stars by making up stories. Then they began to look at stars in a scientific way. And they began to find out the facts that you learned in this book.

Today people explore our solar system in new and better ways. And we are still learning about our solar system. New discoveries are always being made. So you will always be learning new things about our solar system!

- How are we exploring our solar system today?
- What are some new discoveries about the solar system?

You'll learn the answers in this unit.

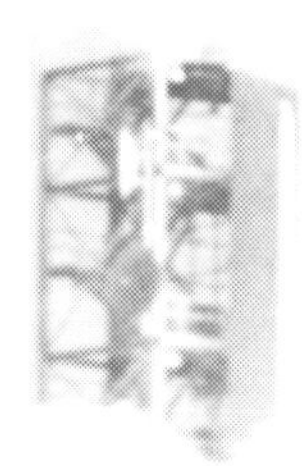

Before You Start

You'll be using the science words below. Find out what they mean. Look them up in the Glossary. On a separate piece of paper, write what the words mean.

1. **robot spacecraft**
2. **space probe**
3. **spacecraft**

NASA photograph

A Giant Leap

"That's one small step for a man, one giant leap for mankind."

That's what **astronaut** Neil Armstrong said when he stepped on the moon in 1969. Neil Armstrong was the first person ever to be on the moon.

Armstrong traveled to the moon in a **spacecraft** called *Apollo 11*. The first person to travel in a spacecraft was Yuri Gagarin, a Russian astronaut. In 1961, he traveled around Earth in the spacecraft *Vostok 1*.

Most of the spacecraft that go into space don't carry any people. They are **robot spacecraft**. Those spacecraft can travel farther than spacecraft with people on them because they can stay in space for a longer time.

Robot spacecraft are sent on space probes—trips through the solar system. They carry sensitive instruments that people on Earth control. Those instruments send information about the solar system back to Earth.

Scientists have sent many robot spacecraft on space probes. *Mariner 10, Viking 1, Voyager 2, and Galileo* are some of those spacecraft. They have helped us make important discoveries about the planets in the solar system.

You may have learned about some of those discoveries through television or the newspapers. Do you remember hearing about any discoveries in space? If so, which ones?

NASA photograph

NASA photograph

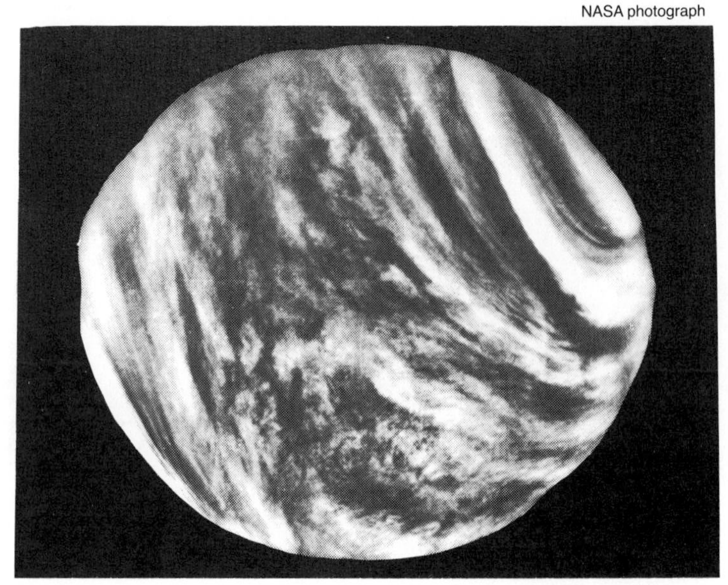

Mercury

Mercury is hard to observe from Earth because it is so close to the sun. In 1974, *Mariner 10* passed close to Mercury. This space probe sent back pictures of the planet. For the first time, scientists could see what Mercury really looks like.

The picture above is one that *Mariner 10* sent back to Earth. It shows part of Mercury. How would you describe this planet?

Like our moon, Mercury has many craters. Most were made billions of years ago. What do you think made the craters on Mercury?

Mercury would not be a comfortable place for humans. During Mercury's long night, it gets colder than any place on Earth. During the day, it's hot enough to melt lead.

Venus

The picture above was also sent by *Mariner 10*. It shows the planet Venus. See the white parts on the planet? What do you think they are?

You're right if you guessed they are clouds.

Until spacecraft visited Venus, scientists thought the clouds were like the clouds on Earth. They thought the clouds were made up of water.

Today scientists know the clouds of Venus aren't like Earth's clouds. They contain a chemical called sulfuric acid. Scientists have also found that the temperature on Venus is very, very hot—about 900 degrees F.

The space probe *Magellan* circled Venus in 1990 and 1991. This probe mapped the surface of Venus. It discovered volcanos, lava flows, and sand dunes.

NASA photograph

Mars

The space probes *Viking 1* and *Viking 2* landed on Mars in 1976. The pictures above were taken by *Viking 1*.

One mission of these probes was to see if there was life on Mars. They took pictures. They tested the soil and air. They could find no sign of life.

But some scientists still think that Mars may have had living things in the past. There are signs that water once flowed on the surface. And there are signs that the atmosphere was once thicker.

Mars would be an exciting place to explore. It has the tallest mountain in the solar system. This mountain is a volcano three times as tall as the tallest mountain on Earth! Mars also has the biggest canyon. It is deep enough to swallow the Earth's tallest mountain.

NASA photograph

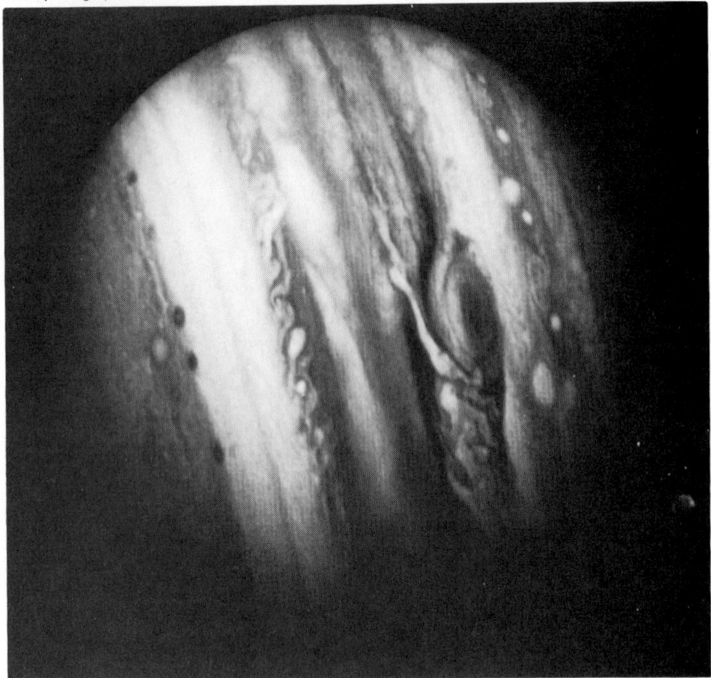

Jupiter

Jupiter is not like Earth, Mars, Venus, and Mercury. It is much bigger and made up mostly of gases. Jupiter has a very thick atmosphere of gases. For these reasons, it is often called a gas giant. Saturn, Neptune, and Uranus are also gas giants.

The center of Jupiter is probably made of very hot rock. This center, or core, is about the size of Earth. Around the core is a layer of chemicals. The chemicals have been squeezed into liquid by the force of the thick atmosphere above.

Look at the picture of Jupiter above. It was taken by the space probe *Voyager 1*. Notice the bands that go around Jupiter. They are made by different kinds of chemicals in the planet's atmosphere.

Voyager 1 discovered that Jupiter has narrow rings high above the clouds of the atmosphere. They are like the rings of Saturn but much smaller. Jupiter also has at least 16 moons.

Scientists saw an amazing thing happen to Jupiter in 1994. Pieces of a comet crashed into the planet. The crashes made huge marks on the planet. Scientists could easily see the marks from Earth.

NASA photograph

Saturn

This picture of Saturn was taken by *Voyager 2*. Notice the rings around the planet. What do you think they are made of?

The rings are made of pieces of dust and ice. Most pieces are small specks. The largest are the size of a house. They revolve around the planet.

Before *Voyager 2* explored Saturn in 1980, scientists thought Saturn had only three rings. Now we know there are many more rings than that. Saturn has seven main rings and thousands of small rings.

Saturn is similar to Jupiter. It probably has the same kind of core. The atmospheres of the two planets are also alike.

Saturn has at least 20 moons. One of these moons, Titan, is close to the size of Mars.

Many people say that Saturn is the most beautiful planet. Why do you think they say that?

NASA photograph

The Faraway Trio

Uranus, Neptune, and Pluto are the farthest planets from the sun.

Uranus and Neptune are about the same size. They are both gas giants with solid cores. Both have rings. Scientists think Uranus may have an ocean of liquid water under its thick atmosphere.

Very little was known about these planets until *Voyager 2* reached them. *Voyager 2* flew by Uranus in 1986. It reached Neptune in 1989.

Pluto was first seen in 1930 through a powerful telescope. It was the last planet to be discovered.

Unlike the gas giants, Pluto is small and solid. Until 1992, scientists thought it was made of rock and ice. Now we know it is made of frozen gases.

Pluto has a moon half its own size. The moon is called Charon.

What Next in Space?

So far, humans have not visited any other planets. But scientists want to send people to Mars someday. It might also be possible for humans to visit Venus. Why can't humans ever land on Jupiter?

Right now, spacecraft carrying people are staying closer to Earth. Spacecraft called space shuttles carry astronauts into orbit around Earth. These spacecraft can then land like airplanes and be used again.

Scientists plan to use the space shuttles to build a large space station. The station would orbit Earth. People could live on the station for long periods. On the station, they could study Earth, the sun, and other planets.

Someday there may be space stations orbiting other planets, too. What do you think it would be like to live on a space station?

Review

Show what you learned in this unit. On a separate piece of paper, answer these questions. Look back through this unit if you need help.

1. Why do scientists use spacecraft without people to explore space?
2. Which planet in our solar system interests you the most? What have you learned about that planet so far?

Check These Out

1. Many satellites made by people revolve around Earth. Find out what some of those satellites are and what they are used for.
2. A spacecraft's trip to the moon is called a *moonshot*. Neil Armstrong traveled on a moonshot in 1969. How many moonshots have carried astronauts to the moon? Find out about one of those moonshots. Then write a story or make a drawing about that moonshot.
3. Imagine that you've been picked to live in a space station. On a separate piece of paper, make a list of six things you would want to take with you. Next to each of those things, write why you would want to take it.
4. Here are more things you may want to find out:
   - What are observatories? How do they help us explore space?
   - What instruments are inside *Voyager 2*? How do they send information back to Earth?
   - How do scientists on Earth control spacecraft that are moving through the solar system?

Show What You Learned

What's the Answer?

Choose the correct endings for these sentences. Each sentence has three correct endings.

1. Our solar system includes
   a. the sun.
   b. planets and moons.
   c. asteroids, comets, and meteoroids.
   d. billions of stars.
2. All planets in the solar system
   a. are made up of hot and burning gases.
   b. rotate around an axis.
   c. have gravity.
   d. revolve around the sun.
3. Earth
   a. is the third planet from the sun.
   b. rotates completely in 24 hours.
   c. is a planet and a star.
   d. revolves around the sun in one Earth year.
4. We have explored our solar system by
   a. sending astronauts to the moon.
   b. sending spacecraft on space probes.
   c. sending astronauts to Mars.
   d. sending people to work in the space shuttle.

What's the Word?

Give the correct word or words for each meaning.

1. A strong pull that all planets and stars have
   G ________
2. Spin around
   R ________
3. A huge ball of very hot gases that gives off heat and light; a sun
   S ________
4. Trips through the solar system to gather information about space
   S ________ P ________
5. The path a thing makes as it moves around another thing
   O ________
6. Billions of stars that form a group in space
   G ________

Congratulations!

You've learned a lot about our solar system. You've learned

- What's in the solar system
- What stars and planets are
- How planets move
- And many other important facts about the planets in our solar system

CHANGING EARTH

How is Earth changing? What is Earth made of? How were mountains and continents formed? How do volcanoes and earthquakes change Earth? How do people change Earth? In this section, you'll learn many facts about changing Earth. And you'll learn how those changes play an important part in our lives.

Contents

Introduction

Do you remember hearing news stories like these?

Volcano Blows in Hawaii!

Snowstorms Bury the Northeast!

Earthquake Shakes California!

Those happenings are news because they can change people's lives. But they can also do this: They can change the surface of Earth.

Ever since Earth began, it has been changing. And it is changing still. Some changes happen suddenly, such as when a volcano blows. But most changes happen very slowly.

In this section, you'll learn how Earth changes. You'll find out what makes those changes. You'll also find out how the land you live on now is changing.

When you finish this section, you'll know about the main kinds of land on Earth. You'll understand how they came to be the way they are. And you'll learn what scientists think will happen to Earth in the years to come.

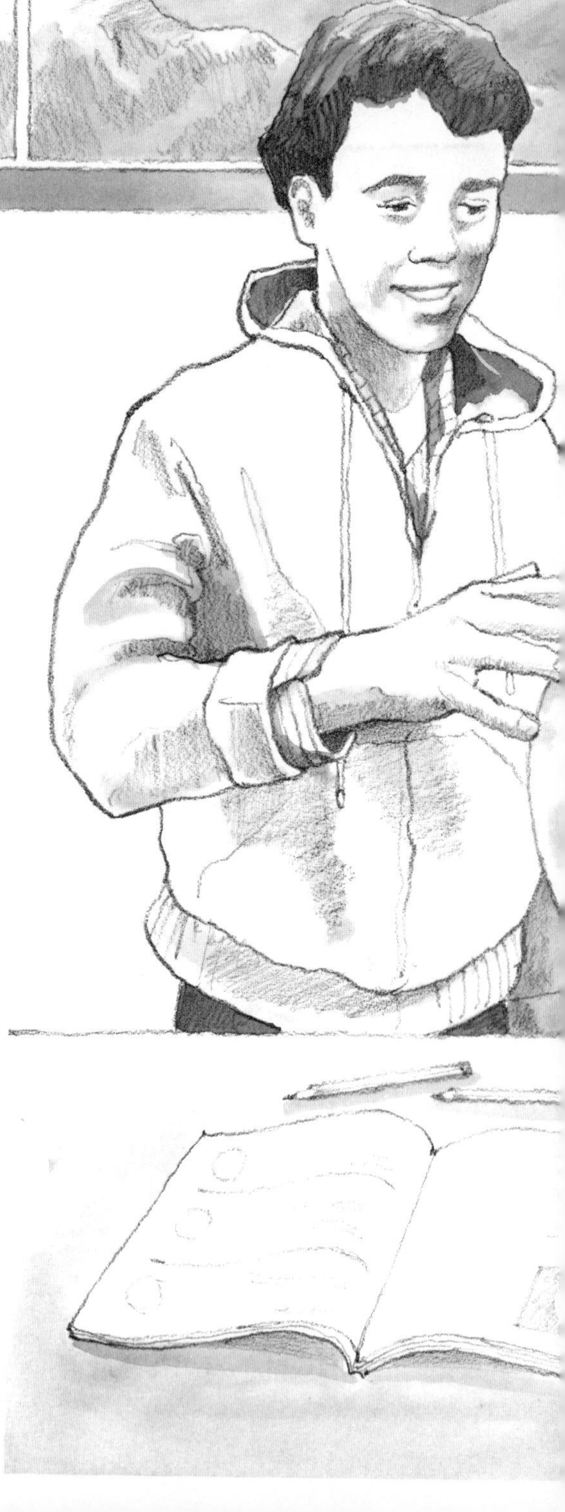

Unit 1

The Land Around You

If you live in Colorado, you might see mountains from your window. But if you live in Kansas, the land around you is probably flat.

Land all over Earth is not the same. Some of it is very flat. Some of it is tall and rocky. It may look as if it always has been flat or tall. But the land has changed. It is changing still.

- What is the surface of Earth made of?
- What are the main shapes of land?
- How is Earth changing?

You'll learn the answers in this unit.

Before You Start

You'll be using the science words below. Find out what they mean. Look them up in the Glossary that's at the back of this book. On a separate piece of paper, write what the words mean.

1. **continent**
2. **evidence**
3. **surface**

Top of the World

You live on the *surface* of Earth. The surface is made up of three parts. One part is solid, one part is liquid, and the other part is made up of gas. You live on the solid part.

What is the solid part called?

Right! It's called land. Land covers all of Earth. We live on pieces of land called *islands* or *continents*.

Look at a map or a globe of the world. It shows some of Earth's land. It also shows the liquid part of Earth's surface. What is that part?

Right! Water is the liquid part of Earth's surface. Water covers nearly three-fourths of Earth. Most of that water is in our oceans.

The last part of Earth's surface is made up of gas. It covers the whole Earth, both the land and the water. What is that part of Earth's surface?

Right! That part is air. Air is also called the **atmosphere**.

Land, water, and atmosphere are all part of Earth's surface. Each of those three parts is always changing and moving. For example, you know that the atmosphere changes. One day it can be sunny and clear. The next day it can rain.

How do you think land can change?

How do you think water can change?

U.S. Geological Survey

The Rocky Mountains are in the middle of the United States.

Landform: Mountains

G.K. Gilbert, U.S. Geological Survey

The Great Plains are in the Midwest.

Landform: Plains

The Shape of the Land

Look at the land around you. What does the land look like? Is it flat? Are there any **mountains** nearby?

What you see is the *shape* of the land. Land can have different shapes. Those shapes are called **landforms**. A hill is an example of a landform. So is a **valley**.

Scientists who study land are called **geologists**. They say there are three main kinds of *landforms*. Those landforms are mountains, **plains**, and **plateaus**.

Mountains are tall landforms. Mountains rise up from the surface. They are often very steep.

Plains are low, flat landforms. They can cover many miles. They are found in the middle of continents. They are also found next to oceans.

Plateaus are landforms that are not as tall as mountains. They are also not as flat and low as plains. They can be hilly. They can have valleys.

The photographs above show the three main kinds of landforms: mountains, plains, and plateaus.

Ross Hall

The Columbia Plateau is in the State of Washington.

Landform: Plateau

Changing Landforms

Mountains look as if they will stay the same forever. But mountains can change. They can break apart. They can wear down. As the mountains wear down, they can change into plateaus.

Those plateaus can also change. They can wear down and change into a landform that is low and flat. What landform is that?

Right! That landform is a plain.

Did you know that plains can also change? They can change into mountains! Geologists think that the Rocky Mountains once were as flat as the Great Plains of the Midwest.

It takes millions of years for one landform to change into another. Geologists study the changes. They look for *evidence*—clues—to prove that landforms have changed.

For example, geologists sometimes find seashells in rocks that are at the tops of mountains. What do you think that tells the geologists about the mountains? (Hint: Where do you usually find seashells?)

Right! At one time, the mountains were at a seashore. They were probably low and flat then.

Geologists also find evidence that Earth is still changing. What do you think some evidence might be?

Earth Watch

Scientists learn about Earth by looking at the world around them. Keep a record of what you learn about changing Earth. On a separate piece of paper, write the answers to these questions.

1. Where do you live?
   (Write the city or town and the state.)
2. Which kind of landform do you live on?
   What does it look like?

Review

Use what you learned in this unit to answer these questions.

1. Which three are landforms?
   a. Mountains
   b. Plateaus
   c. Oceans
   d. Plains
   e. Air
2. What two ways can landforms change?
   a. Mountains can break apart and wear down.
   b. Flat lands can become mountains.
   c. The atmosphere changes into mountains.

Check These Out

1. Make a Science Notebook for this section. Use it to keep a record of what you learn about changing Earth. Put your list of glossary words and their meanings in the notebook. Also keep your notes from the Earth Watch sections in it. You can keep anything else you learn about changing Earth in your Science Notebook too.
2. Get a globe that shows you the landforms of Earth. (It will show you how high and low those landforms are.) Turn the globe so that you are looking at Antarctica (the land around the South Pole). Is there more land or more water on that side of the world? Find all the continents.
3. Scientists think Earth is about 4½ billion years old. They call this amount of time *geologic time*. Find out more about geologic time. When did life begin on Earth? What are the time periods?
4. Find out what these other landforms look like. Where are they found?
   - Beaches
   - Islands
   - Caves
   - Drumlins
   - Mesas
   - Canyons

Unit 2

Inside Moves

Suppose you work in a store. You come in one morning and see this: Shelves tipped over. Boxes and broken bottles on the floor. Cracks in the wall. What happened was an earthquake!

You probably felt that earthquake. An earthquake is one example of how Earth can move. Geologists think that the continents weren't always where they are today. They think the continents must have moved from somewhere else.

- What is Earth like beneath the surface?
- What causes Earth to move?
- What causes earthquakes?

You'll learn the answers in this unit.

Before You Start

You'll be using the science words below. Find out what they mean. Look them up in the Glossary. On a separate piece of paper, write what the words mean.

1. **fault**
2. **fossil**
3. **layer**

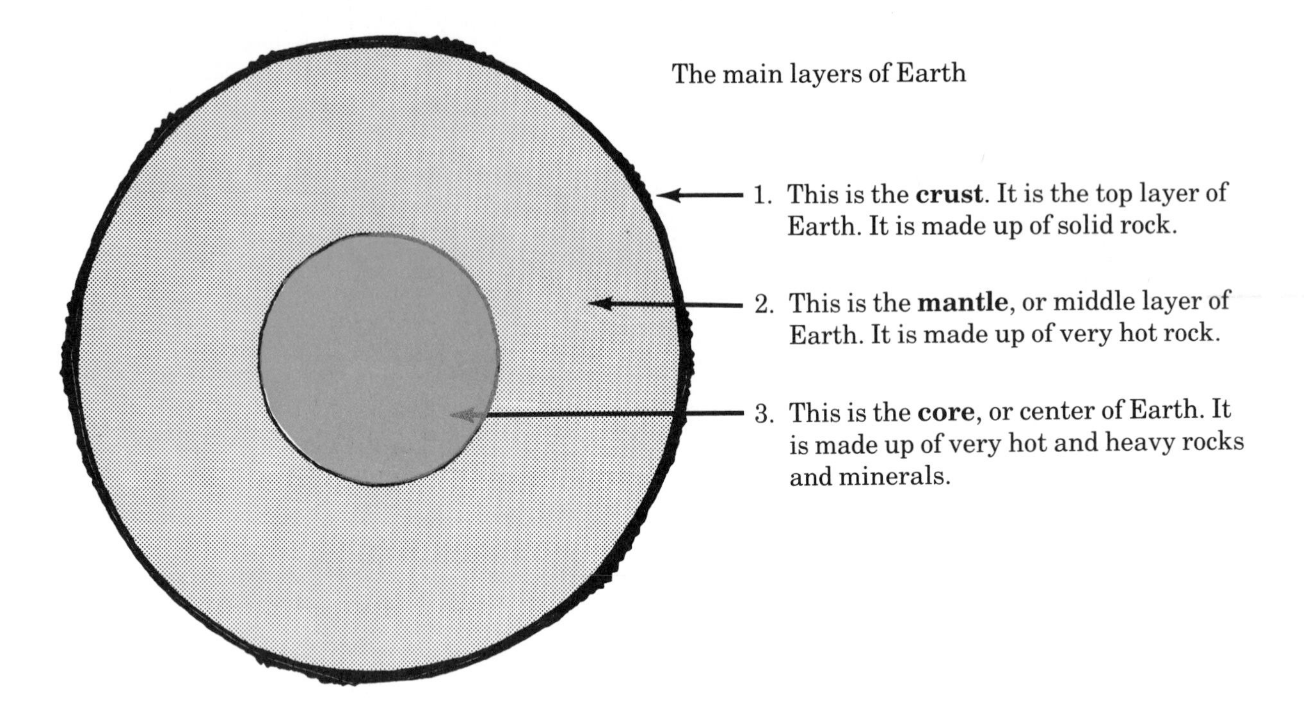

Beneath the Surface

Suppose you could dig a hole to the center of Earth. What do you think you would find?

You would find different *layers* of rocks.

No one has ever been able to dig to the center of Earth. (In fact, no one has been able to dig through the top layer of Earth.) But geologists think they know what the different layers of rocks are like. The diagram on this page shows the main layers.

The top layer is called the **crust**. It is made up of solid rock. It is the thinnest layer of Earth.

The middle layer is called the **mantle**. Geologists say that the mantle is made up of very hot rock. Some of the mantle is partly melted.

The last layer makes up the very center of Earth. Geologists think that layer is made up of very heavy rocks and minerals. They also think it is very hot. What do you suppose that layer is called? (Hint: Think of the center of an apple.)

Right! The center of Earth is called the **core**.

Henry Degenkolb, U.S. Geological Survey

In 1906, San Francisco was almost destroyed by an earthquake.

R.E. Wallace, U.S. Geological Survey

The San Andreas Fault caused the San Francisco Earthquake.

Moving Rock

Have you ever been in an earthquake? If you have, what was it like?

The ground probably shook hard. Chairs and lamps jumped around. Maybe parts of buildings fell down. All that happened because the Earth's crust moved. It shook everything on top of it.

Sometimes the crust moves a lot. That's what happened in 1906 when a terrible earthquake shook San Francisco.

Geologists believe that the San Francisco earthquake was caused by a *fault* near the city. A fault is a long deep crack in the Earth's crust.

How does a fault move? Put your hands together so your palms touch. Each of your hands is like the side of a fault. Now slowly move your hands past each other. The sides of a fault can move past each other like that. When that happens, the crust shakes—and there's an earthquake.

Earthquakes happen somewhere on Earth every day. Many of them are so small we don't feel them. But geologists know they happen because they use **seismographs**—machines that can record even the smallest earthquakes.

Seismographs show that many earthquakes happen in the western United States. Where else in the world do they happen?

Moving Faults

You can see the ways faults move by making a clay model. You'll need these materials:

- Four pieces of clay, each a different color
- One ruler

You can use the modeling clay that toy stores sell. Or you can make baker's clay:

Mix 2 cups of flour with 1 cup of salt. Add about 1 cup of water to make a stiff clay. Knead the clay until you can stretch it.

Divide the clay into four pieces. Color three of the pieces. Make each one a different color. Put a few drops of food coloring or poster paint in each piece. Knead the clay until the color is mixed in.

---

1 Flatten the four pieces of clay with your hands. Put them on top of each other to make a stack of different-colored layers. The layers are like layers of different kinds of rocks in Earth's crust.

Use the edge of the ruler to make a line across the top of the clay. That line is like a road on Earth's crust.

1

2 Cut across the line. Then cut the stack in half. You now have a crack in the clay. What is that crack like?

Right! The crack is like a fault in the crust.

2

3 Pick up the two halves of clay. Move one half up. Keep the other half down. That's one way a fault can move. What happens to the layers in each half?

Right! The layers in one half don't match the layers in the other. Geologists tell how a fault has moved by looking at layers of rocks in the two sides of a fault. They see if the layers match.

3

4 Now place the two halves of clay on your table. Put them together again. Make sure the lines in the two halves match.

5 Move the two halves past each other. That's another way a fault moves. What happens to the "road" on top of the clay?

Right! It breaks into two lines.

Geologists look for things that should be in a straight line, such as roads. If roads or layers of rocks are in two parts, they know that a fault has moved.

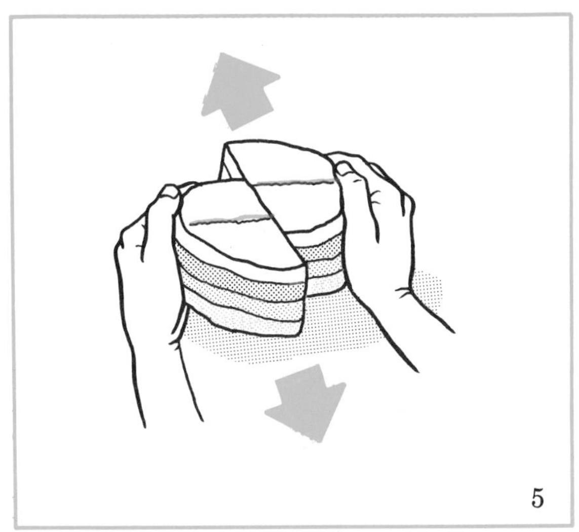

6 Geologists can also measure how much the fault has moved. How can they do that?

They measure the distance between the two parts.

Get a ruler. Put it on the crack at the top of the clay. Measure the distance between the two lines. How far did your fault move?

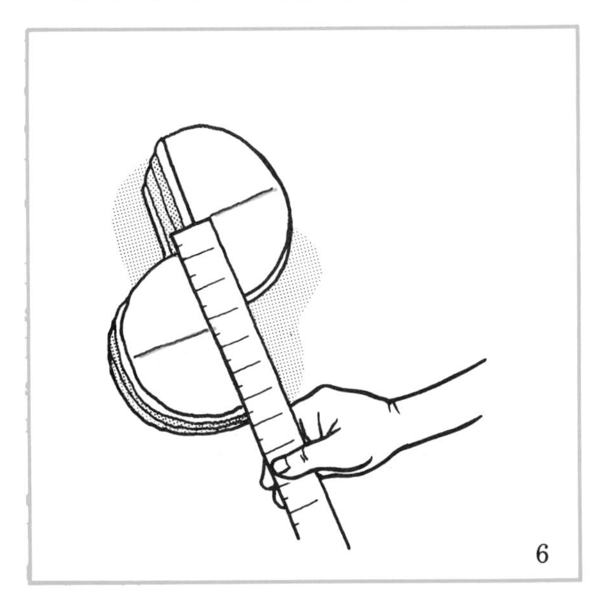

Earth today has many continents.

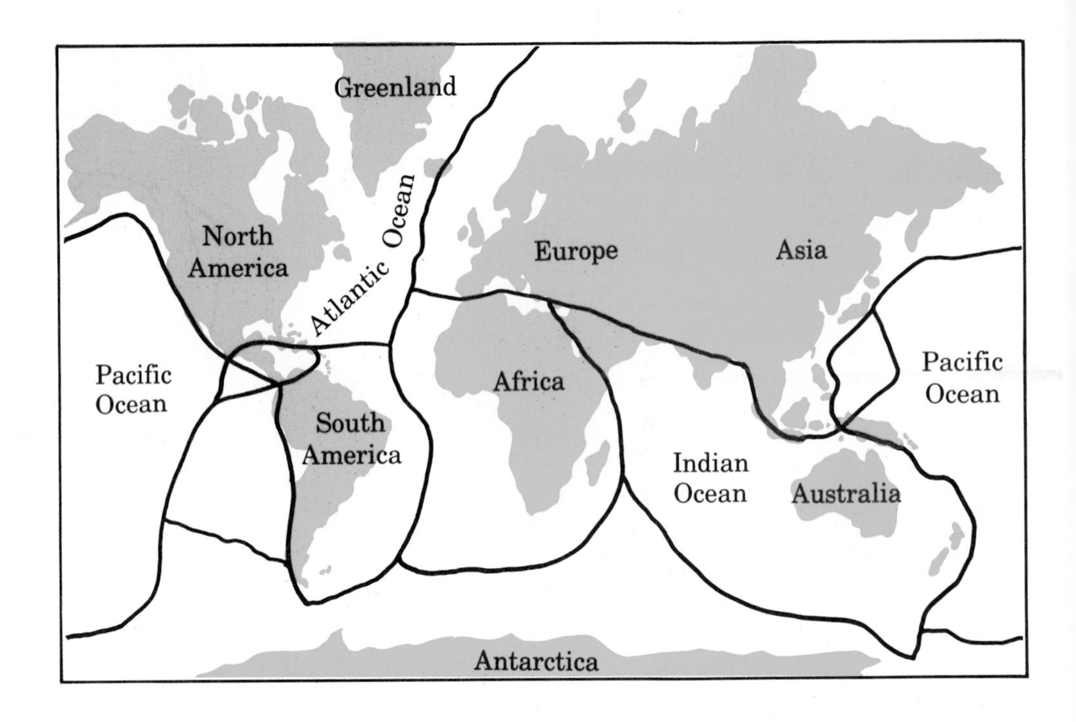

Pieces of Crust

You learned that a fault is a long, deep crack. The crust all over Earth is cracked by faults. The map on this page shows Earth's main faults. Notice this: They break up Earth's crust into huge pieces.

Geologists call those pieces **plates**. They say that the plates are really huge blocks of solid rock. On the map, find the plate that your country is on. What continent is on the plate?

What ocean is on the plate?

Geologists think that the plates ride on Earth's middle layer, the mantle. (Remember: Some of the mantle is made up of partly melted rock.) They believe the plates are always moving.

Of course, the plates move very slowly. But they can move as much as two inches (about five centimeters) a year. That may not seem like much, but in a million years those inches will add up!

When a plate moves, the land on top of it moves too. For example, geologists say that part of California is slowly moving north. Where do you think that part will be a million years from now?

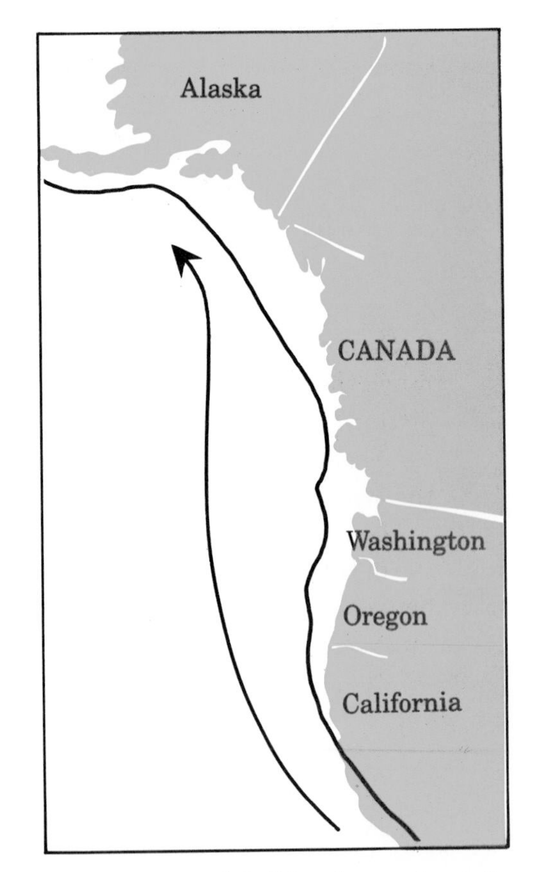

Where is part of California moving to?

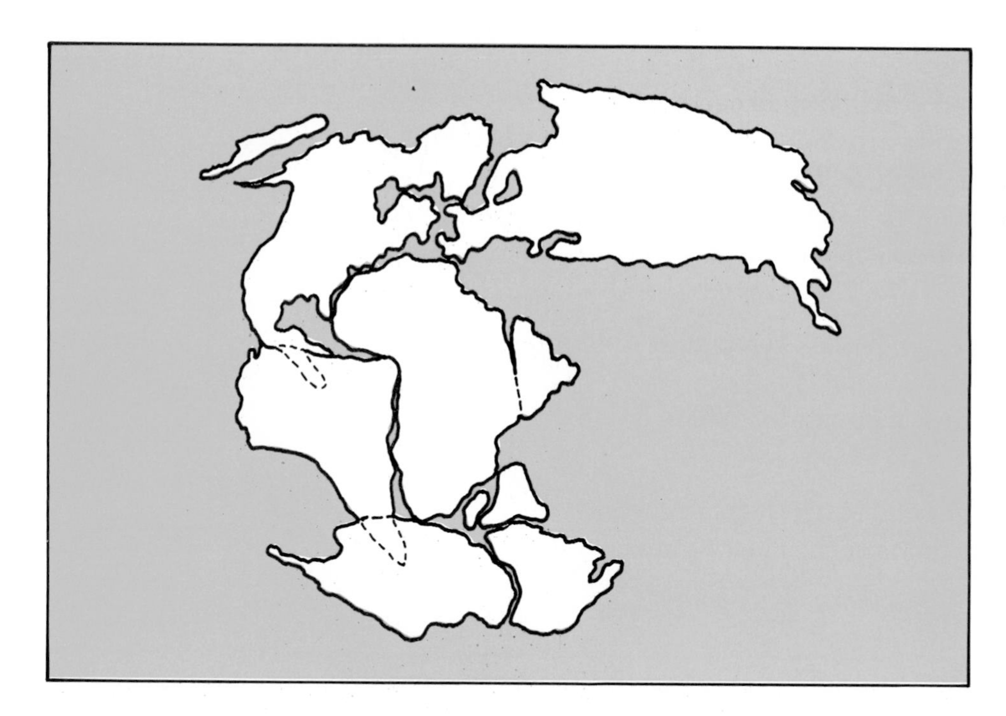

Geologists believe that Earth once had only one huge continent.

Moving Continents

What do you think Earth looked like 200 million years ago? The map on this page shows what many geologists think. They think Earth had only one huge continent, surrounded by one huge ocean.

Look at the map on page 62. It shows Earth today. Earth now has seven continents. What do you think happened?

The huge continent was probably split by faults. They split the continent into seven pieces that slowly moved away.

Geologists think that's what happened because of certain evidence. Here's one piece of evidence: Look again at the map of Earth today. Look at the shapes of the continents. Geologists say those shapes are like the pieces of a jigsaw puzzle. The shapes look as if they could all fit together to form one huge piece.

Here's more evidence: All the continents have certain kinds of rocks. And sometimes they have fossils of plants and animals that lived at the same time. Those rocks and fossils show geologists that the continents were once the same land.

Fossils also show geologists where a land might have moved from. For example, Greenland is close to the North Pole. It is very cold. But scientists have found fossils on Greenland that show it once was very warm. Where do you think Greenland moved from?

Earth Watch

Find out when your state has had earthquakes. Find out where the faults are. Your state government has that information. Write to *The Geological Survey Office* at your state capital. Ask these questions. On a separate piece of paper, keep a record of what you find out.

1. Have there been earthquakes in our state? Where? When?
2. Are there any faults in our state? Where is one located?

Get a map of your state. Mark the places where earthquakes happened. Mark the places where faults are.

Review

Show what you learned in this unit. Match the layers listed below with the correct clues.

crust mantle core

1. This layer is cracked by faults.
2. Plates ride on top of this layer.
3. This is the center of Earth.
4. During an earthquake, this layer moves.
5. These two layers are hot rock.

Check These Out

1. Watch newspapers and magazines for stories about earthquakes. Find out where the earthquakes happened. Mark those places on a map of the world. How big were the earthquakes? Keep the articles you find in your Science Notebook.
2. Find out more about the earthquakes that happened in these cities. Which earthquake caused the most damage?
   - Lisbon, Portugal, 1755
   - New Madrid, Missouri, 1811–12
   - San Francisco, California, 1906
   - Anchorage, Alaska, 1964
   - Mexico City, Mexico, 1985
3. Put tracing paper on a large wall map. Trace the outlines of Africa, North America, and South America. Cut them out. Then try to put them together like the pieces of a jigsaw puzzle. How well do they fit?
4. Project Mohole was a plan to drill a hole through Earth's crust. Find out if scientists were able to finish it.

Unit 3

Mountains

Did you ever climb to the top of a mountain? If you did, you may have noticed how that mountain seemed to be pushed up from the surface of Earth. You may have looked down and seen other mountains. Maybe you wondered how mountains are formed.

Mountains are formed from changes in Earth. Sometimes the changes take thousands of years. Sometimes they take just a few years.

- What changes in Earth make mountains?
- How are mountains formed?

You'll learn the answers in this unit.

Before You Start

You'll be using the science words below. Find out what they mean. Look them up in the Glossary. On a separate piece of paper, write what the words mean.

1. **collide**
2. **magma**
3. **pressure**

A. Keith, U.S. Geological Survey

The Appalachian Mountains, eastern North America

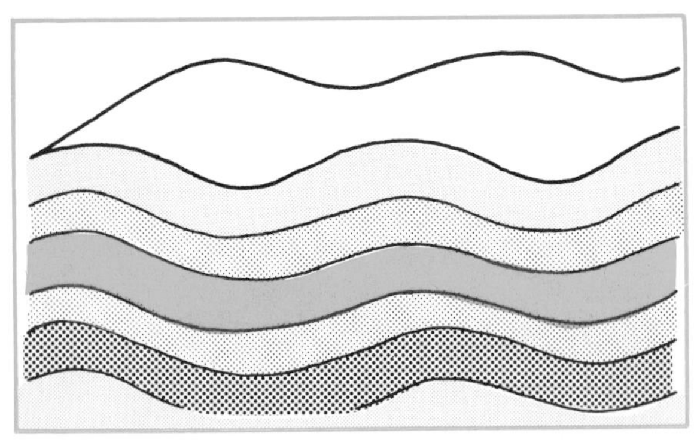
Folded mountains

Grand Teton National Park

The Grand Teton Mountains, Grand Teton National Park, Wyoming

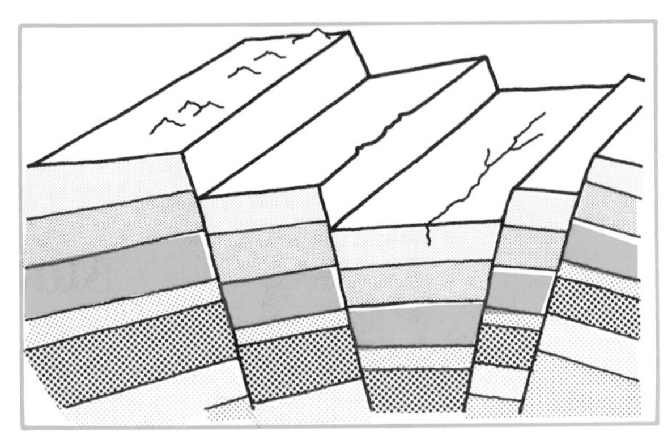
Block mountains

Pushing Up Mountains

You learned that Earth's plates can move away from each other. But Earth's plates can also move toward each other. They can *collide* with each other and slowly push up new landforms. What are those landforms?

Right! The landforms are mountains. (Remember that plates are huge blocks of Earth's crust and they ride on the mantle. Part of the mantle is made up of *magma*—melted rock.)

Sometimes, when plates collide, this happens: The two plates slowly push against each other. That causes the rocks on their edges to become wrinkled and folded. Then those rocks get pushed up. They become mountains.

You can see the folds and wrinkles on those mountains when you look at them from an airplane. The Appalachian Mountains were formed this way.

Another thing that can happen is this: When two plates collide, one slides under the other. The plate on top gets slowly pushed up. The faults on the crust break apart and large blocks of rocks are formed. Those blocks get slowly pushed up. They form mountains.

You can see the blocks when you look at those mountains from far away. The Grand Teton Mountains were formed in this way.

Look at the photographs at the top of this page. Which one shows mountains that were formed by the crust folding up? Which one shows mountains that were formed by one block being pushed up?

Make Two Mountains

Make models of two kinds of mountains. You will need these things:

- Two balls of modeling clay (or baker's clay)
- Two pieces of cardboard

Knead the clay until it is soft and stretches. Then follow the directions on this page.

Folded Mountain

1 Divide one ball of clay in half. Flatten the two pieces so they are about ½ inch thick. Those pieces are like the plates of Earth's crust. Place the two pieces next to each other on one of the pieces of cardboard.

2 Push the two pieces together very slowly. The clay should wrinkle and fold. As you push, the clay should rise up to form a folded mountain.

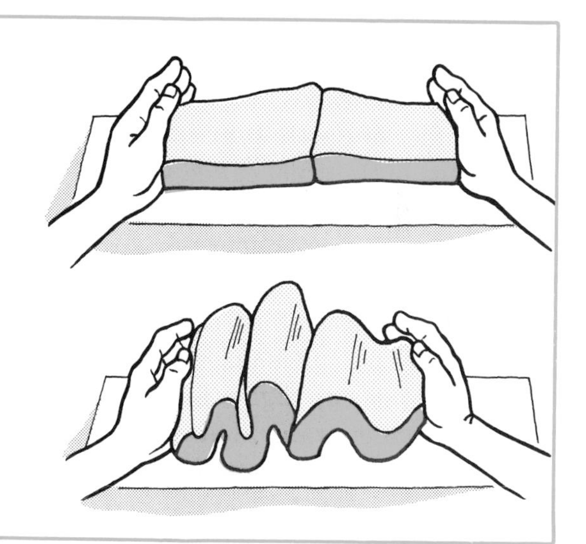

Block Mountain

1 Divide the other ball of clay in half. Flatten the two pieces so they are about ½ inch thick. Place the two pieces next to each other on the other piece of cardboard. Put the edge of one piece over the other.

2 Push the two pieces together so one slides under the other. The top piece should rise up to form a block mountain.

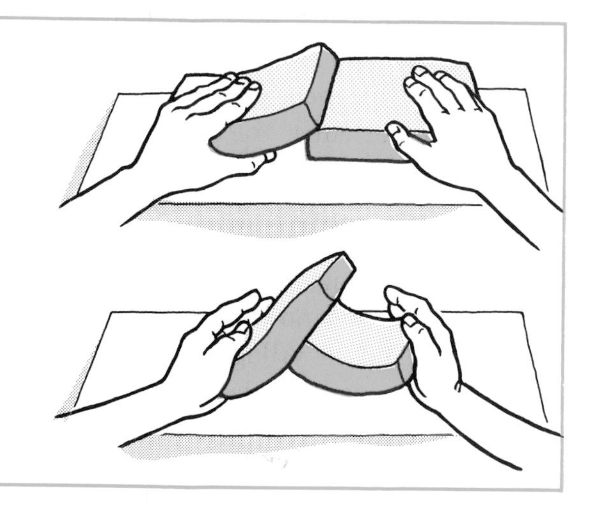

Now make labels for your two models. Write *block mountain* on one, and *folded mountain* on the other. Let the models dry. Then paint them to look like real mountains.

Volcanoes

You know that there is very hot, melted rock deep inside Earth. When melted rock is inside Earth, we call it magma. Sometimes magma flows up to the surface of Earth. It **erupts** through openings in the crust called **vents**. Then we call it *lava*.

Lava can form a mountain. We call that kind of mountain a volcano. How do you think lava can form a mountain?

Right! Lava piles up each time it erupts. It piles up higher and higher to build up a mountain. Mount St. Helens is a volcano in Washington that was built that way.

It can take thousands of years to build a volcano. For example, it took 40,000 years to build up Mount St. Helens. But sometimes volcanoes can rise up in only a few years. Paricutin Volcano in Mexico is an example.

In 1943, lava suddenly erupted in the middle of a flat cornfield close to a village. That was the start of Paricutin Volcano. By 1952, the volcano was more than 1,300 feet high. And the village was gone— completely covered by lava.

Volcanoes can also form islands. One of our states is made up of islands that were formed by volcanoes. In fact, those islands are the tops of volcanoes. What is the name of those islands?

Right! The Hawaiian Islands are the tops of volcanoes. Those volcanoes built up from the ocean floor.

Austin Post, U.S. Geological Survey

Mount St. Helens erupted in 1980.

W.F. Foshag, U.S. Geological Survey

K. Segerstrom, U.S. Geological Survey

Paricutin Volcano grew out of a flat cornfield.

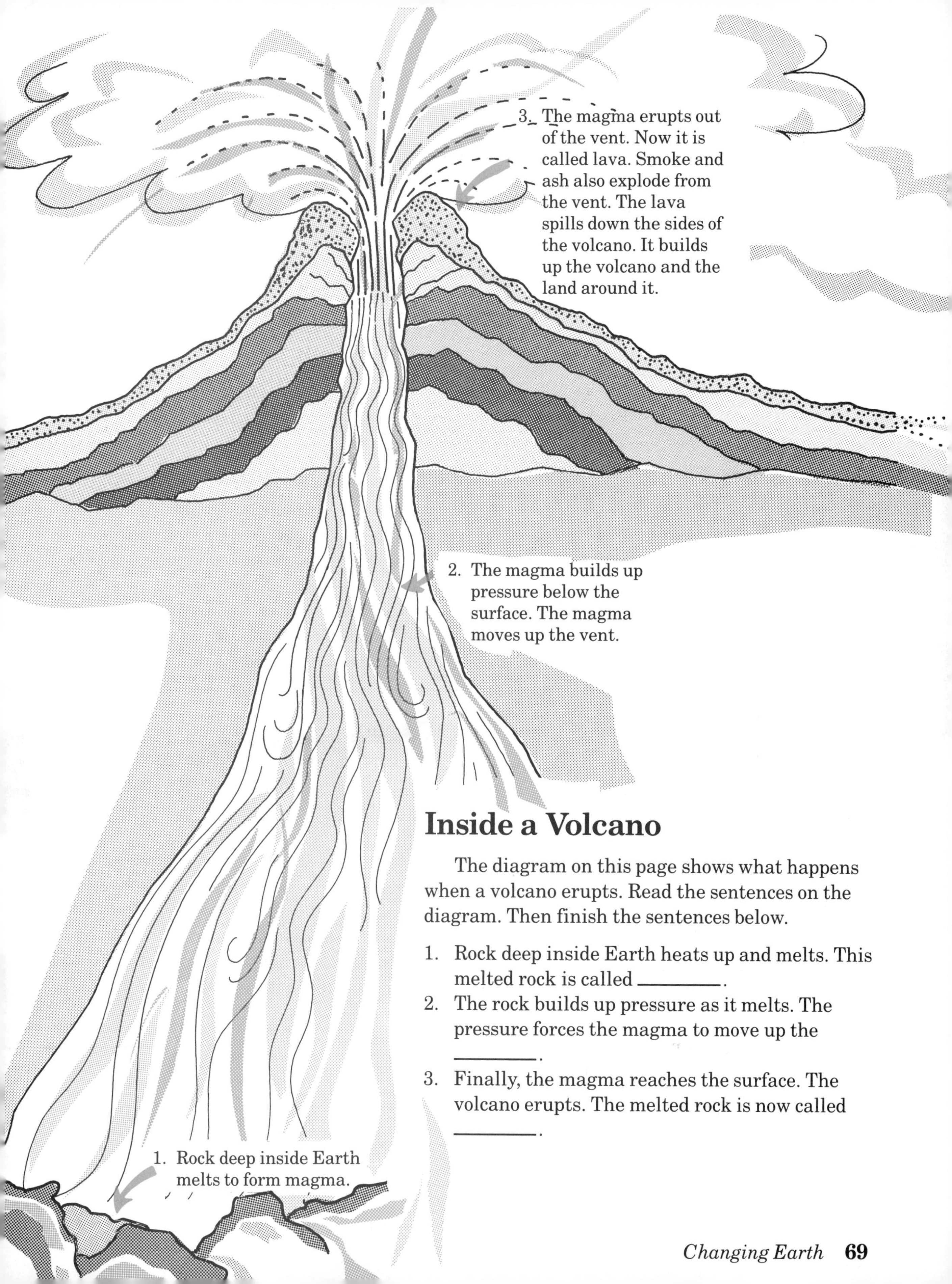

Inside a Volcano

The diagram on this page shows what happens when a volcano erupts. Read the sentences on the diagram. Then finish the sentences below.

1. Rock deep inside Earth heats up and melts. This melted rock is called ________.
2. The rock builds up pressure as it melts. The pressure forces the magma to move up the ________.
3. Finally, the magma reaches the surface. The volcano erupts. The melted rock is now called ________.

Earth Watch

Are there any mountains in your state or in a nearby state? Find out where the closest mountain is. On a separate piece of paper, keep a record of what you find out.

1. What is the name of the mountain?
2. What kind of mountain is it: folded, block, or volcano?

Review

Show what you learned in this unit. Find the right words to finish the sentences.

1. Rock on the crust forms two kinds of mountains when
   a. plates of Earth's crust collide.
   b. plates move from each other.
   c. plates don't change.
2. Folded mountains are formed
   a. when a volcano erupts.
   b. during an earthquake.
   c. when plates collide and push up rock.
3. Block mountains are made
   a. from very hot lava.
   b. when plates push against each other and fold up.
   c. when one plate slides under another.
4. Volcanoes are mountains that build up
   a. when rocks break apart.
   b. when melted rock flows from deep inside Earth to the surface.
   c. when mountains become plateaus.

Check These Out

1. Mount St. Helens is just one of the volcanoes in the United States. Find out what other volcanoes are in this country. When did they last erupt? Which one may erupt next?
2. Which is the highest mountain in the United States? In the world? Find out. Look in an encyclopedia under *mountains*.
3. Watch the newspapers for articles about volcanoes. Add these articles to your Science Notebook. Mark the volcanoes' locations on your map.
4. Find out more about the Appalachian Mountains and the people who live there.
5. Make a poster that shows the three main kinds of mountains.
6. How do scientists tell when a volcano might erupt? Find out and report to the class.

Unit 4

Wearing Down Mountains

As hard as a rock. That's what we say when we want to describe something that's so tough it can't be changed.

But even the hardest rock can be changed. In fact, rocks are always changing. They can become different kinds of rocks. They can be broken down. They can make new landforms.

- How do rocks change?
- What breaks up rock?
- How do changing rocks make new landforms?

You'll learn the answers in this unit.

Before You Start

You'll be using the science words below. Find out what they mean. Look them up in the Glossary. On a separate piece of paper, write what the words mean.

1. **dissolve**
2. **erode**
3. **sediment**

Changing Rocks

You know that Earth's crust is made of solid rock. That rock is always changing.

Rock can be changed by the heat deep inside Earth. The heat melts the rocks inside the crust. The rocks become magma. The magma then moves up toward the surface. It cools and hardens into another kind of rock. We call rock that forms that way **igneous rock**.

Rocks on top of Earth's surface can also change into another kind of rock. Here's how: The rocks are broken up into tiny bits—*sediment*. The sediment gets into oceans and lakes. It forms layers on the bottoms of the oceans and lakes. The layers harden into rock. That new rock is called **sedimentary rock**.

Rocks deep inside Earth can also change into another kind of rock. The tremendous heat and pressure inside Earth causes the rocks to change. The heat and pressure make the rocks very hard. We call rock that changes that way **metamorphic rock**. A diamond is a metamorphic rock.

Look at the photographs of the different rocks on this page. Answer these questions. Then check your answers. (The right answers are upside down.)

1. What kind of rocks are *granite* and *basalt*?
2. What kind of rocks are *sandstone* and *shale*?
3. What kind of rocks are *gneiss* and *slate*?

Wilberg/Gee

These rocks were formed from hot melted magma or lava.

Wilberg/Gee

These rocks were formed from layers of sediment.

All other photos by U.S. Geological Survey

These rocks were formed by heat and pressure deep inside Earth.

Answers

1. igneous 2. sedimentary
3. metamorphic

Breaking Them Up

Rocks are usually hard and strong when they are first formed. But as time passes, things happen that can weaken the rocks and break them up. What do you think could happen to break up rocks?

Trees and other plants can grow on rocks. They start growing in small cracks. They send out roots. What happens as a plant grows bigger?

Right! As a plant grows, its roots grow thicker and longer. The roots force the crack to open wider. If the roots grow big enough, they can split the rock apart.

Weather can also help break up rocks. Rain and melting snow fill cracks in rocks with water. When the water freezes, it takes up more space. The ice pushes the sides of the crack apart. This happens over and over again as the water melts and freezes. The cracks get wider and wider and the rock breaks apart.

Wind helps break up rock, too. Hard winds can pick up sand and carry it. The sand gets blown against rocks. The sand particles scrape away tiny bits of rock.

Some rocks can be weakened in another way. As rainwater seeps through dead plant material in the soil, it picks up certain chemicals. The chemicals make the water a weak acid. This acid seeps into the rocks below. If those rocks are limestone or marble, the acid can slowly *dissolve* them. Tiny bits of the rock are slowly carried away. Cracks get wider as a result.

Wind can blow sand against rock. These rocks have been scraped by sand in the wind.

Running water breaks up large rocks into smaller ones.

Rock Breaker

Water can get into cracks in rocks. When the temperature drops below 32°F (or 0°C), that water freezes into ice. Ice can break up rocks. Find out why. You will need these things:

- One small empty milk carton
- Water
- One stapler
- Tape
- A freezer

1 First, fill the milk carton all the way to the top with water. Then staple across the top of the carton about five times. The top should be tightly shut. Tape over the top.

2 Place the carton in the freezer. Leave it there until the water turns to ice—about 24 hours. What happens to the carton? On a separate piece of paper, draw a picture of the carton.

What Happens?

1. What happens to water when it freezes?
2. Water **expands** (takes up more room) when it freezes. What can happen when water freezes in a crack in a rock?

Another Rock Breaker

What happens when acid rain falls on rocks such as limestone or marble? Find out! You will need these materials:

- One felt pen
- One cup of water
- One cup of an acid, such as vinegar
- Two small glass jars
- Two small pieces of limestone

1 Write *water* on one jar. Pour 1 cup of plain water into that jar. Write *acid* on the other jar. Pour 1 cup of vinegar into that jar.

2 Put a piece of limestone in each jar. Put the jars in a safe place. The next day, look at the limestone in each jar.

What Happens?

1. What happens to the limestone in the plain water?
2. What happens to the limestone in the acid?

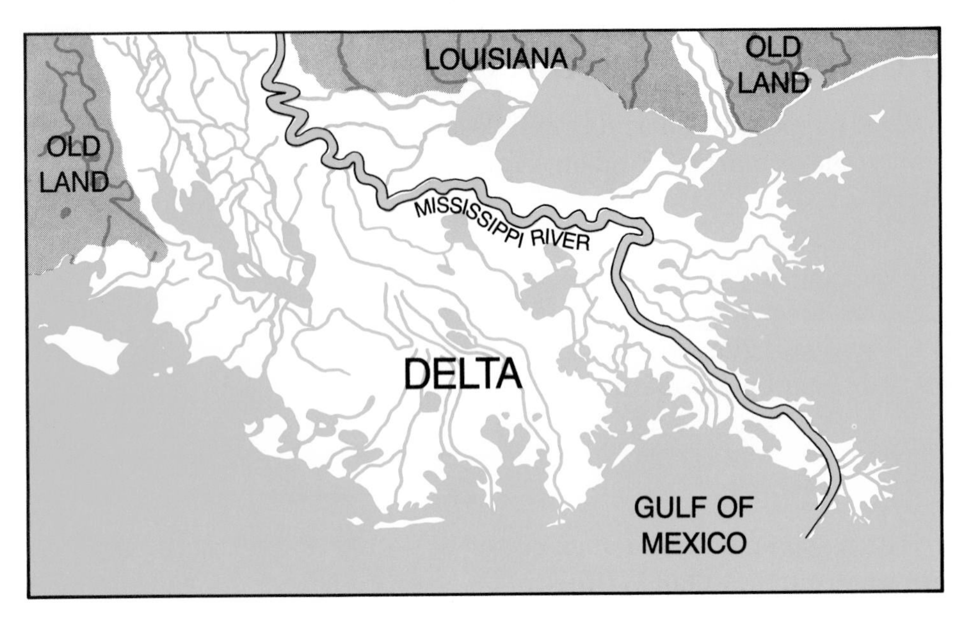

This map shows the Mississippi Delta. The Mississippi flows into the sea here. The fan-shaped landform is the delta.

Take Them Away

Rocks get broken into small pieces. What do you think happens to those pieces?

Rain carries away those pieces.

Rainwater can wash sediment down hills and mountains into streams. Small streams then carry that sediment into large streams. The large streams carry the sediment into rivers. The rivers then carry the sediment into a lake or ocean.

Water must move in order to carry sediment. The faster the water moves, the bigger the pieces of sediment it carries. What do you think happens to the sediment when the water slows down?

Right! The sediment is **deposited**. It drops to the bottom. The biggest pieces drop first. As the water slows down, the smaller pieces drop. When the water stops completely, all the pieces slowly drop out.

Sediment that is deposited this way can build islands in the middle of rivers. It can form *deltas* where a river runs into an ocean. A delta is a landform that is shaped like a fan. Deltas can spread for many miles.

Look at the map. It shows where the Mississippi River runs into the Gulf of Mexico. Find the delta on the map.

N.H. Darton, U.S. Geological Survey

1. Part of the Great Plains, South Dakota

G.S. Rogers, U.S. Geological Survey

2. Garden of the Gods, Colorado

Yosemite National Park

3. Yosemite Valley, Yosemite National Park, California

Erosion

Suppose a landform is very flat. Then water and wind begin to break up the rocks on that land. Rain carries away sediment. What happens to that landform?

Right! The landform *erodes*. It wears down.

The photograph on the top of this page shows one part of the Great Plains. The land used to be very flat. Now it has many *gullies*. Gullies are big cuts in the land that are made by running water. The water breaks up rocks. It washes away the rocks and erodes the land.

Erosion can wear down a landform. The Great Plains are an example. They are being eroded by running water.

Wind can also cause erosion. Strong winds can pick up pieces of sand and blow them against rocks. The sand rubs off bits of the rocks the way sandpaper rubs off bits of wood.

There is another way landforms can be eroded. Landforms can be eroded by **glaciers**. What are glaciers made of?

Right! Glaciers are made of ice.

Glaciers have many huge rocks frozen to them. As the glaciers move, those rocks scrape the land. They dig out land and form huge valleys.

Now look at the pictures. They show landforms that are eroded. Tell what caused the erosion: ice, wind, or water? Then check your answers. (The right answers are upside down.)

Answers

1. water 2. wind 3. ice

Make a Stream Table

A stream table shows how water causes erosion. You make a model of a landform. Then you spray water on it. The running water makes gullies. It also makes deltas. To make a stream table you need these materials:

- One large, flat plastic or foil pan
- Scissors or knife
- Sand
- One shower hose
- Waterproof tape
- Water faucet
- Two buckets

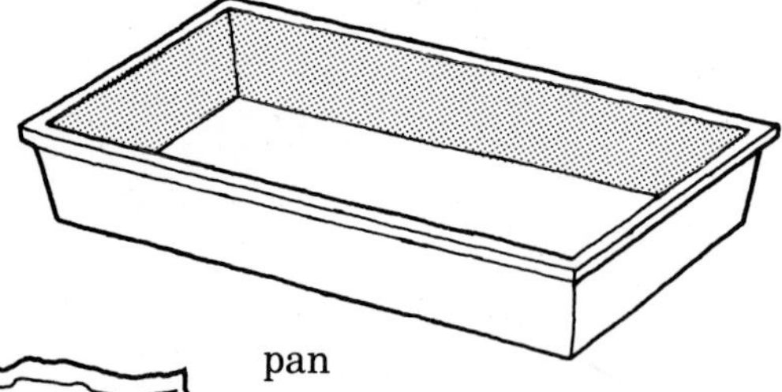

pan

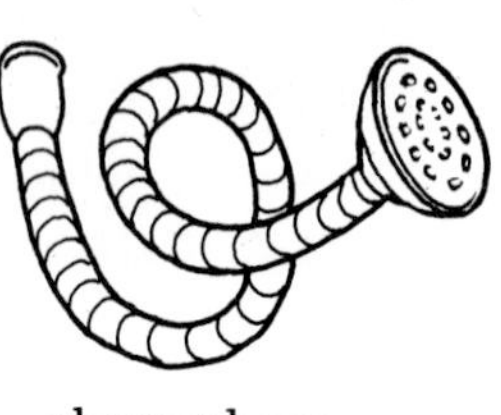

shower hose

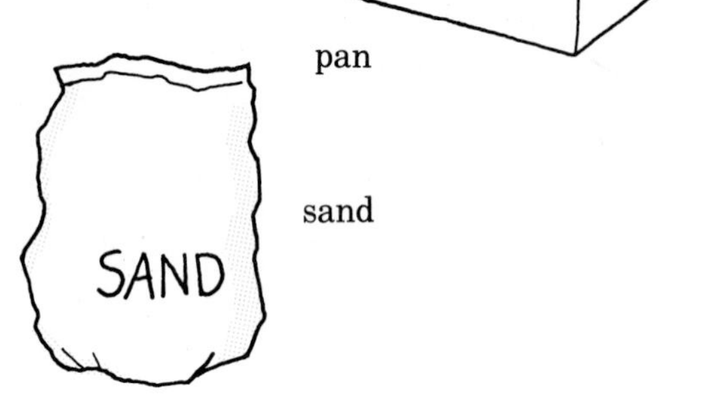

sand

---

1 Cut a hole in one end of the pan. Cut it close to the bottom.

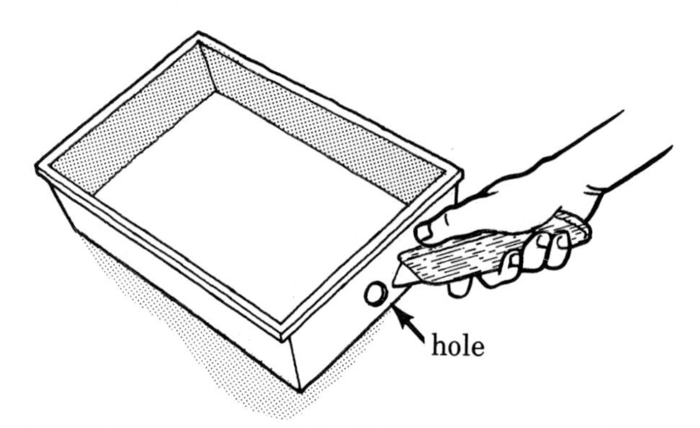

2 Place the pan on a counter or table near the water faucet. Put the end with the hole a little over the edge.

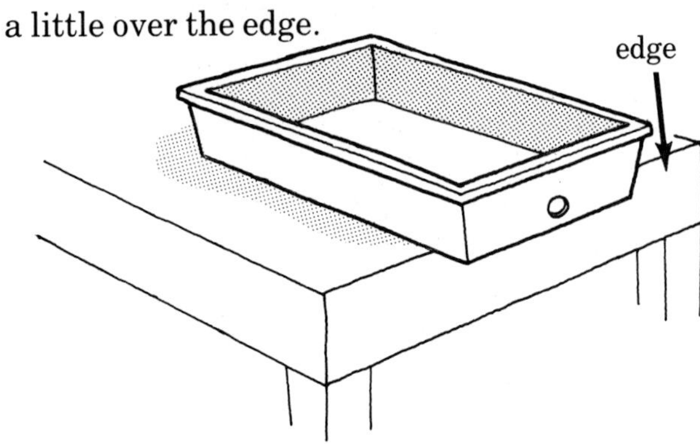

---

3 Pour sand into the pan at the end that doesn't have the hole. Make a little hill with the sand.

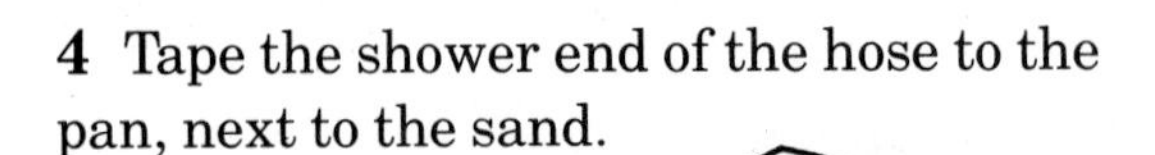

4 Tape the shower end of the hose to the pan, next to the sand.

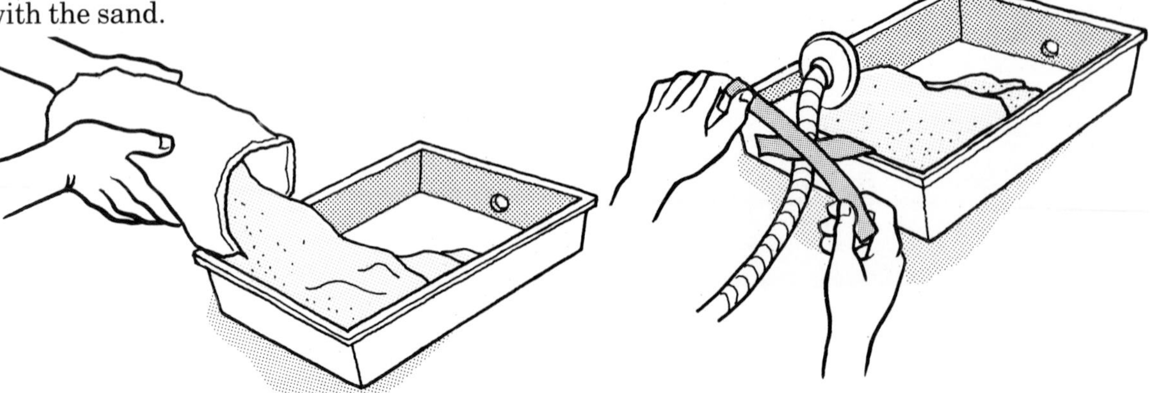

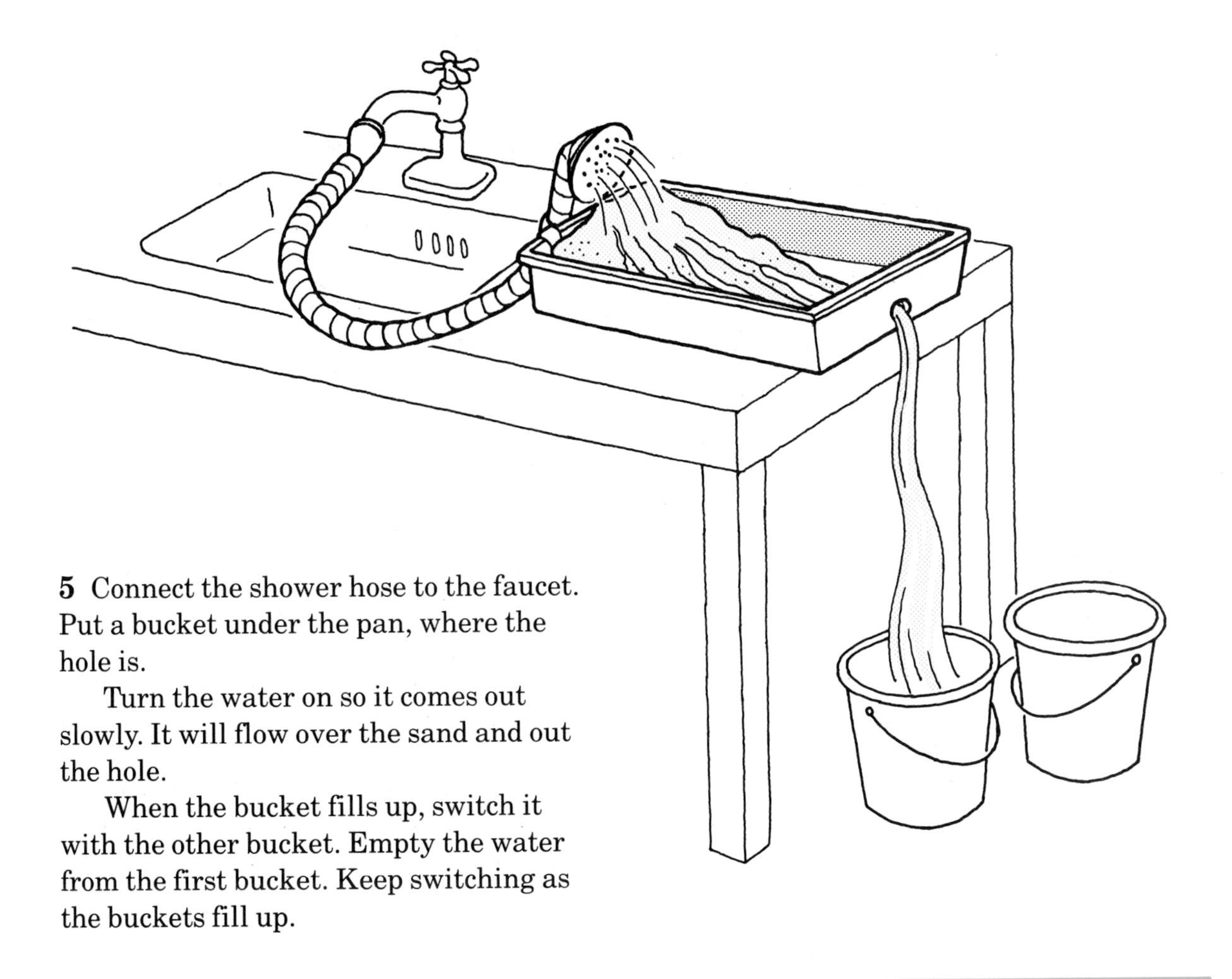

5 Connect the shower hose to the faucet. Put a bucket under the pan, where the hole is.

Turn the water on so it comes out slowly. It will flow over the sand and out the hole.

When the bucket fills up, switch it with the other bucket. Empty the water from the first bucket. Keep switching as the buckets fill up.

6 Watch what the water does. Does it form streams? Does it make little lakes? Watch how it makes gullies in the hill and carries away the sand.

After 20 minutes, turn off the water. What happens to the sand? Look for landforms such as plateaus, plains, deltas, and islands. On a separate piece of paper, draw a picture of what the sand looks like now.

Earth Watch

Locate a big river on a map of your state. Trace over the river with a light-colored marking pen. Use the map to answer the questions below. On a separate piece of paper, keep a record of what you find out.

1. What is the name of the river?
2. Where does it start?
3. What does it flow into?

Review

Use what you learned in this unit to answer the questions. Then check your answers. The page where you'll find the answer is listed after each question.

1. What are some ways that rocks change? (page 72)
2. What causes rocks to break up? (page 73)
3. What happens to rock that is broken up? (page 76)

Check These Out

1. Gems are found in many different kinds of rocks. Look up these gems in an encyclopedia. Find out how they were formed and where they are found.
   - Garnet
   - Tourmaline
   - Ruby
   - Amethyst
   - Diamond
2. The Grand Canyon was eroded by a river. What river is it? How long did it take to erode the canyon? Find out.
3. Fill a peanut butter jar halfway with dirt. Then fill it the rest of the way with water. Put the lid on it. Shake the jar hard. Let the jar sit overnight. What does the dirt look like now? On a separate piece of paper, draw a picture of it.
4. Add a glacier (ice cube) to your stream table. What happens as it melts?
5. Some great floods have changed the lives of many people. Find out about these floods:
   - Johnstown, Pennsylvania, 1889
   - Rapid City, South Dakota, 1972
   - Big Thompson Canyon, Colorado, 1976

Unit 5

Climate

Think about a cold winter day. People wear heavy coats. Snow covers the ground. An icy wind blows across the land.

Now think about a hot summer day. The air seems wet and heavy. People wear light clothing. The sun seems to burn the land.

Cold winters and hot summers are certain kinds of *climate*. Earth has many different kinds of climate. A climate can be cold, warm, or hot. It can also be dry or wet. But no matter what a climate is like, it causes changes on Earth.

- How do climates change Earth?
- How do scientists study those changes?

You'll learn the answers in this unit.

Before You Start

You'll be using the science words below. Find out what they mean. Look them up in the Glossary. On a separate piece of paper, write what the words mean.

1. **ice age**
2. **moraine**

Susan Kaschner Jagoda

Death Valley National Monument is in southern California.

Hawaii Visitors Bureau

Hawaii is one of the Hawaiian Islands.

Cool or Warm, Wet or Dry

Death Valley is a desert. What kind of climate do you think Death Valley has?

Hawaii is an island in the middle of the Pacific Ocean. What kind of climate do you think Hawaii has?

Right! Death Valley is *hot* and *dry*. And Hawaii is *warm* and *wet*.

Hot, warm, dry, wet. Those are the kinds of words we use to describe the climate of a place. Climate is the kind of weather a place has year after year.

The climate in one place can be very different from the climates in other places. For example, Seattle, Washington, is usually wet and cool. The climate of Phoenix, Arizona, is usually warm and dry.

The climate a place has can help change the land. How do you think the land can change where the climate is very rainy and wet?

Right! Where there is lots of rain, running water can erode the hills. Plants grow faster and their roots help break down rocks.

Sometimes a landform causes the climate to be a certain way. For example, mountains sometimes cause rain to fall. The rain forms when wind pushes wet air up the side of the mountains. That rain usually falls just on one side of the mountains.

Which side of the mountains do you think erodes faster?

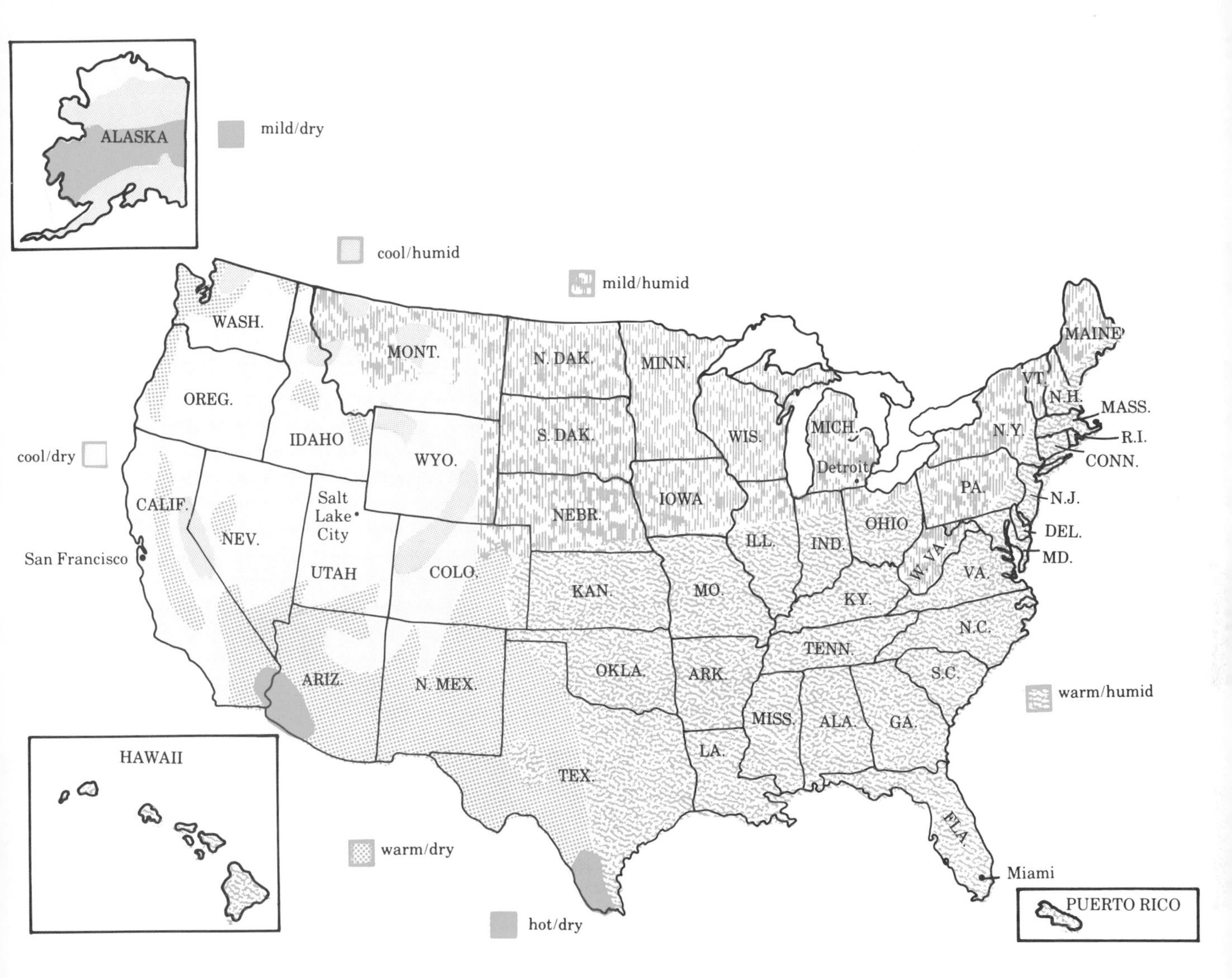

What's Your Climate?

Some places don't have the same climate all year long. For example, they may have summers that are warm and **humid** (warm/humid). They may also have winters that are cold and wet (cold/wet). The map shows the main regions of the United States. It also shows the summer climate for those regions. Each shade stands for a certain climate.

Look at the map. Find the cities listed in the next column. Tell what kind of climate each city has during the summer. Then check your answers. (The right answers are upside down.)

1. San Francisco, California
2. Salt Lake City, Utah
3. Detroit, Michigan
4. Miami, Florida

Find the region you live in on the map. What kind of summer climate does your community have?

Answers

1. cool and dry
2. cool and dry
3. mild and humid
4. warm and humid

Climates of the Past

One hundred million years ago, Earth was warm and humid. Tropical plants grew in huge forests. Then, about 70 million years ago, the climate changed. Earth started cooling and the tropical plants died.

We know that happened because of certain evidence that scientists have found. The evidence gives us clues about what climates were like long ago.

To find evidence, scientists drill deep holes in the ocean bottom. They take out **core samples**. A core sample shows the layers of sediment that make up the ocean bottom. Scientists study the fossils in a layer to see how hot or cold a climate was. Scientists measure how thick the layer is to see how long that climate lasted.

Scientists also study the bottoms of lakes. They look for layers of sediment that were left when the glaciers melted. If they find a very thick layer of sediment, that tells them the glaciers were melting fast. What do you think the climate was like to cause that?

Right! The climate was very warm.

Scientists also look for evidence in trees. Trees grow a new layer of wood every year. We call those layers *growth rings*. You can see the rings when the tree trunk is cut across.

By studying growth rings, scientists can tell what the climates were like many years ago. When the climate is wet and warm, a tree grows fast and its ring is wide. When the climate is very dry, a tree grows slowly, and its ring is thin.

The photograph shows the growth rings of a tree. Find the rings that were formed when the climate was very wet.

U.S. Geological Survey

The arrows mark every tenth year of the tree's life. Notice that the arrows are farther apart during the wet years than during the dry years. What does that tell you about how fast or slowly the tree grew?

H.E. Malde, U.S. Geological Survey

Saskatchewan Glacier is in the Columbia Ice Field, Alberta, Canada.

Yosemite National Park

Half Dome in Yosemite National Park was formed by glaciers.

Ice Ages

Earth has had several *ice ages*. During the ice ages, the climate was very cold for millions of years. Sheets of ice, some a mile thick, covered large parts of many continents. Scientists believe the last ice age began about two or three million years ago.

During the ice ages, glaciers covered most of North America. They started as small ice sheets in northern Canada. As the ice sheets grew, they started to move.

The glaciers picked up huge rocks. They carried the rocks for hundreds of miles. They scraped and dug up land. They carved huge valleys as deep as Yosemite Valley in California. They hollowed out lakes as large as the Great Lakes in the Midwest. They pushed rocks and soil ahead of them. They ground up rocks into a fine dust.

Then the climate of Earth changed. It grew warmer. The glaciers slowly melted. The rocks and dirt in the glaciers were deposited. Streams of water from the glaciers carried some sediment far away. Other sediment was dumped at the glaciers' edges. That sediment formed large hills called *moraines*. You can find moraines in Michigan and in many other states in the Midwest.

Some scientists think that the last ice age isn't over yet. They think the glaciers might come back. What do you think would happen if the glaciers did come back?

Earth Watch

On a separate piece of paper, answer the questions below.

1. What is the summer weather like where you live?
2. What is the winter weather like where you live?
3. Check with your parents, grandparents, and older friends to find out about any strange weather that might have happened where you live. For example, were there any very cold and snowy winters?

Review

Show what you learned in this unit. Then check your answers. The page where you'll find the answer is listed after each question.

1. Give an example of a climate. (page 83)
2. How can a very wet climate change the land? (page 82)
3. How can glaciers change the land? (page 85)
4. How do scientists find out about past climates? (page 84)

Check These Out

1. Find out what the chinook, Santa Ana, and sirocco winds are. Where do they happen?
2. *Climatologists* are scientists who study climate. Find out what they do. Find out about other jobs that have to do with climate.
3. What parts of our country did the glaciers cover during the last ice age? Did they cover the place where you live? On a separate piece of paper, draw a map that shows what parts were covered by glaciers.
4. Scientists have several ideas about why the dinosaurs disappeared. Some think they died out because the climate changed. Find out what the climate was like when the dinosaurs lived. What do scientists say changed the climate?

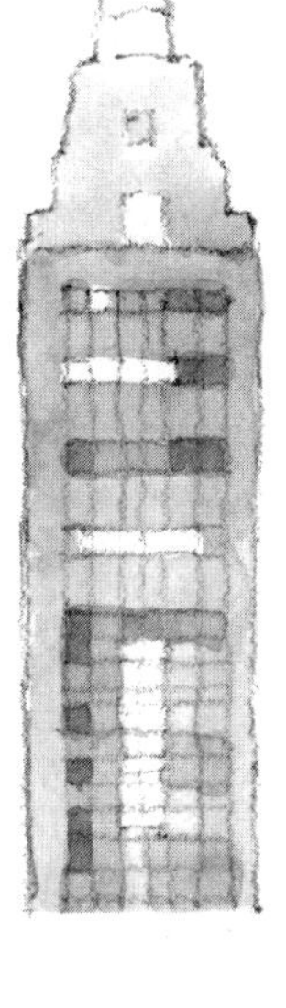
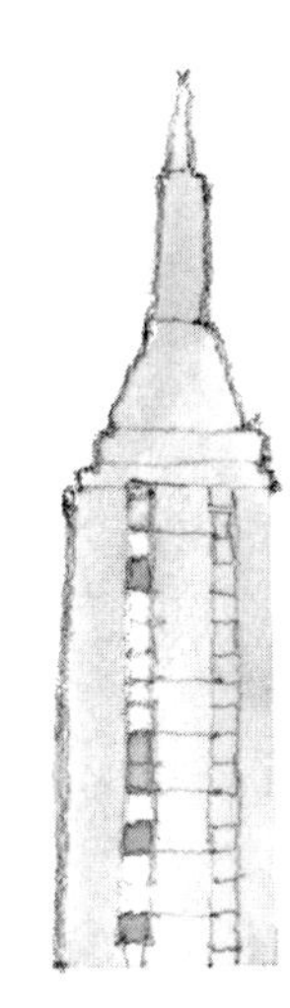

Unit 6

People Change the Earth

You've learned some ways that Earth changes. The crust moves and changes the shape of a land. Volcanoes erupt and build up mountains. Wind, rain, and ice break down rocks and make new landforms. All those things happen because of changes in Earth's crust and in the atmosphere.

But Earth is also changing because of *people*. Before people came, Earth's changes sometimes took thousands of years to happen. The changes that people make can happen very quickly.

- What do people do that causes changes?
- What are the changes people are making?
- What are people doing about those changes?

You'll learn the answers in this unit.

Before You Start

You'll be using the science words below. Find out what they mean. Look them up in the Glossary. On a separate piece of paper, write what the words mean.

1. **fossil fuels**
2. **greenhouse effect**

J.E. Kinsley, U.S. Bureau of Reclamation

People built this dam across a river. The dam holds back some of the river water to make a lake.

Builders and Makers

The first human beings appeared on Earth about two million years ago. In order to survive, they needed water, food, shelter, clothing, and tools. They got those things from Earth.

Early humans broke up rocks to make tools. They learned to make things from wood and metals. They learned to dig wells for water and to plant crops. As they did those things, they began to change Earth.

Today millions more people live on Earth. We need and want more than early people did. We are changing Earth faster than those early humans ever could.

We fill in rivers and swamps to make new land. We blow up landforms to make roads and bridges. We make lakes to hold water for farming and drinking. We dig deep into the ground for fuel, such as oil, coal, and gas.

We also cover large parts of the land with cities and towns. Here's what happens when we build a new community: First we dig up landforms. Then we cover the ground with concrete. We build buildings. We build streets and parking lots. To make those things, we use rock, minerals, and fuels that we dig from the ground. We use wood from trees we cut down. How do you think all of that changes Earth?

The Greenhouse Effect

Suppose it's a cold day. You get into a car that's been parked in the sun. All the windows have been closed tightly. Inside the car, it's warm! Why is this so?

Light from the sun passes through the glass windows. The sunlight heats the seats and dashboard. They give off heat. The closed windows keep the heat from leaving. So the car stays warm inside.

Earth's atmosphere acts something like the glass windows. Sunlight passes through the atmosphere and heats the land and water. This heat leaves the surface and goes into the atmosphere. Some of the heat keeps going back into space. But the atmosphere traps much of the heat. In this way, the atmosphere keeps Earth warm enough for living things.

Many scientists are worried that changes in the atmosphere will keep Earth too warm. Humans burn lots of fossil fuels like oil and coal. We burn these fuels when we use cars and make electricity. But burning them creates the gas carbon dioxide. The carbon dioxide goes into the atmosphere. It makes the atmosphere trap more heat than before.

Scientists call the possible warming of Earth the **greenhouse effect**. A greenhouse is a glass building for growing plants. If extra heat can't escape from a greenhouse, it can become too hot for the plants inside.

How might Earth change if it becomes warmer?

The atmosphere naturally traps heat. More carbon dioxide can make it trap extra heat.

Looking at the Future

People have been changing Earth for more than a million years. At first, people didn't think about the future. For example, they would cut down all the trees in an area for firewood. They didn't know what that would do to the land.

When all the trees in a large area are cut down, the land erodes quickly. The climate can become dry and windy. After many years, the land can become so rocky that plants won't grow.

People have also burned lots of coal and oil for a long time. They didn't realize the changes they were causing to the land and air. What changes did they cause?

To get coal and oil, people dug mines and wells. That caused erosion of the land. When people burned those fuels, gas and smoke polluted the atmosphere. Chemicals polluted the water and land.

Today people try to figure out what might happen in the future. They study the land before they build on it or dig into it. They try not to ruin the land.

People also know that landforms change. They know what can happen to lands that are near faults and volcanoes, or to low lands that are next to rivers. So before people build on an area of land, they find out how that land might change. Why is that a good idea?

R.E. Wallace, U.S. Geological Survey

This photograph was taken from an airplane. It shows houses that were built next to a fault. Do you think people planned for the future? Why or why not?

Earth Watch

People are probably changing the land near where you live. Think of a place near your home where people are changing the land. It might be a highway that's being built, or a building that's going up. On a separate piece of paper, answer the questions below.

1. What is the name of the place?
2. What is being built there?
3. What changes are happening to the land?

Review

Use what you learned in this unit to answer the questions. Then check your answers. The page where you'll find the answer is listed after each question.

1. How are people changing the land? (page 88)
2. How are people changing the atmosphere? (page 89)
3. How can we plan for the future? (page 90)

Check These Out

1. Suppose you are stranded on a desert island. What do you need to survive? How will you get those things? Write a story to describe how you survive.
2. Find out how fossil fuels are formed. Where do we find large supplies of those fuels in the world today?
3. Find out more about how greenhouses work. How warm can they get? What kinds of plants grow best in greenhouses?
4. The Environmental Protection Agency (EPA) is a government office. It looks after the country's *natural resources.* Find out about the EPA.
5. Find newspaper articles that warn about changes in Earth. Cut out those articles. Make a poster.

G. K. Gilbert, U.S. Geological Survey

Ross Hall

Unit 7

The Land You Live On

You have gathered a lot of information about the place where you live. Look over the Earth Watch pages at the end of each unit. Then, on a separate piece of paper, answer these questions.

1. What kind of landform do you live on? Describe it.
2. What faults are nearest you?
3. What mountains are nearest you?
4. What rivers are nearest you?
5. What kind of climate does your area have in the winter? Spring? Summer? Fall?
6. What kinds of changes are people making in your area?
7. What do you think the land will be like one million years from now? On a separate piece of paper, write a story or draw a picture to show what you think the land will be like.
8. Why do you think the land will be like that one million years from now?

ıaiian Visitors Bureau

A. Keith, U.S. Geological Survey

Show What You Learned

What's the Answer?

Choose the right endings for these sentences. There may be more than one correct ending for a sentence.

1. Three main kinds of landforms are
   a. plains, plateaus, and mountains.
   b. streams, lakes, and oceans.
   c. land, water, and air.
2. Earth's crust is
   a. cracked by faults.
   b. in the center of Earth.
   c. moving.
3. Mountains form when
   a. landforms erode.
   b. plates collide.
   c. volcanoes erupt.
4. Rocks can be worn down by
   a. plants.
   b. acids.
   c. wind and running water.
5. Landforms can be changed by
   a. glaciers.
   b. heavy rain.
   c. growth rings.

Why should we be worried about how people change Earth?

What's the Word?

Give the correct word for each meaning.

1. A large piece of land
   C ________
2. Hot melted rock inside Earth
   M ________
3. To wear away land
   E ________
4. The shape of a piece of land
   L ________
5. A huge, moving body of ice
   G ________
6. Bits of rock
   S ________
7. A scientist who studies Earth
   G ________

Congratulations!

You've learned a lot about our changing Earth. You've learned

- What Earth is made of
- How mountains and continents are formed
- How Earth is changing
- How people can change Earth
- And many other important facts about the planet you live on

---

WEATHER

How does weather change? How do we predict weather? Where does rain come from? What can we learn by looking at clouds? What causes winds? In this section, you'll learn many facts about weather. And you'll learn how weather plays an important part in our lives.

Contents

Introduction

Picture this:

It's a warm summer day. You're getting ready to go to a game. You're not sure of what to wear. The sun is shining, but there are some dark clouds in the sky. You decide to wear a T-shirt and jeans.

Halfway through the game, the weather changes. It starts to rain *hard*! You get soaking wet.

How can you know if weather will change? You can use weather reports. And you can look for signs in the weather that show it may change.

This section will help you learn to do those things. You'll learn

- what makes weather change
- how to look for signs that weather may change
- how to guess what the weather will be
- how to understand weather reports

You'll also learn how to make your own weather reports. And you'll do some experiments that show why some kinds of weather happen.

When you finish this section, you'll be ready for any change in weather—rain or shine!

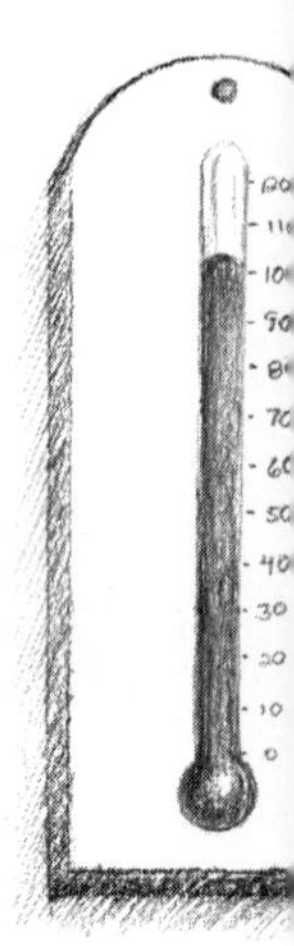

Unit 1

Hot or Cold?

What's the weather like today? Hot? Cold? Warm? Cool?

All those words tell how much heat is in the air. If there's a lot of heat in the air, we feel hot. If there's not much heat in the air, we feel cold.

- How much heat is in the air?
- How does the weather change when the heat in the air changes?

You'll learn the answers in this unit.

Before You Start

You'll be using the science words below. Find out what they mean. Look them up in the Glossary that's at the back of this book. On a separate piece of paper, write what the words mean.

1. **degrees**
2. **measure**

What's the Temperature?

Imagine this:

You're listening to a weather report. You hear the reporter say:

"It's 100 degrees right now."

What's the reporter talking about?

Right! The reporter is talking about how hot the day is. The reporter is telling people how much heat is in the air. In other words, the reporter is giving the **temperature** of the air.

Air temperature is measured by special **thermometers**. The picture shows one kind of thermometer. Notice the numbers on the thermometer. They show the degree, or amount, of heat in the air.

We use those numbers when we talk about temperature. If the weather is hot, the numbers are high. If the weather is cold, the numbers are low.

For example, the temperature on a hot summer day might be 100 degrees. But the temperature on a cold winter day might be 10 degrees. What do you think the temperature might be on a cool spring day?

Changing Temperatures

Suppose you hear a weather reporter say:

"Today's high will be in the 80s."

What is the reporter telling you?

The reporter is telling you what the highest temperature will be during the day. It will be somewhere between 80 degrees and 89 degrees. For example, it may be 82 degrees. Or it may be 88 degrees.

Suppose the reporter also says:

"Today's low will be in the 60s."

What do you think that means?

Right! The lowest temperature during the day will be somewhere between 60 degrees and 69 degrees. It may be 62 degrees, or 66 degrees, and so on.

When the reporter tells you the highs and lows of the day, he or she is giving you an idea of what kind of weather to expect. The reporter is telling you how the air temperature may change. Often, when the air temperature changes, the weather changes too.

What will be the highest temperature in your area today? What will be the lowest temperature? Find out. Listen to a weather report.

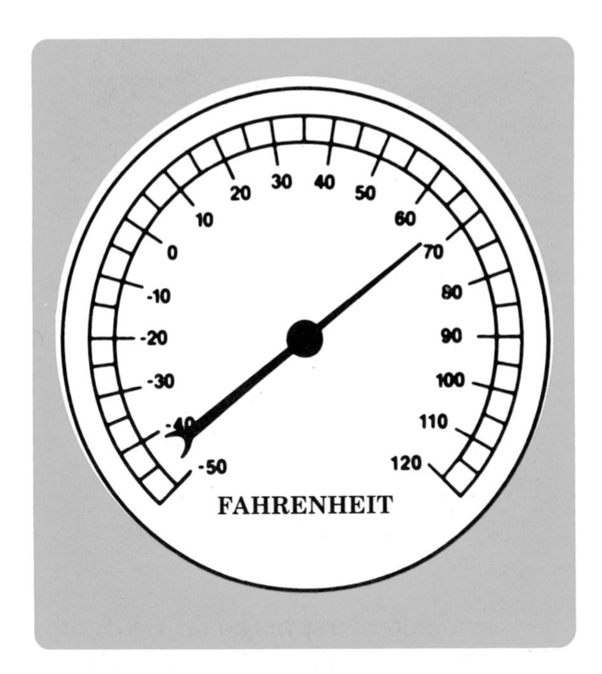

Another kind of air thermometer

Weather Watch

Watch the weather for a day. See how the air temperature changes. See if the weather changes.

Hang an outdoor thermometer outside in a shady place. Put it where the sun can't shine on it.

Look at the thermometer three times during the day: once in the early morning, once around noon, and once in the late afternoon.

On a separate piece of paper, keep a record of the three temperatures. First, write the day's date. Then, each time you look at the thermometer, write these things:

1. Time (Example: 8:30 a.m.)
2. Temperature (Example: 69 degrees)

When you are finished, answer these questions:

1. Look at the temperatures you wrote. What is the highest temperature?
2. What is the lowest temperature?
3. Did the weather change during the day? If it did, describe what happened.

Review

Show what you learned in this unit. Match the words in the list below with the correct clues.

temperature	degrees	thermometer
high	low	measure

1. Find out the amount of something, such as heat
2. Amounts of heat in the air
3. Something that shows how much heat is in the air
4. Heat in the air
5. Highest temperature
6. Lowest temperature

Check These Out

1. Make a Science Notebook for this section. Use it to keep a record of what you learn about weather. Put your list of glossary words and their meanings in your notebook. Also keep your notes from experiments and the Weather Watch sections in it. You can keep anything else you learn about weather in your Science Notebook too.
2. On a separate piece of paper, draw a Fahrenheit thermometer. Write these degrees on it: 110, 90, 70, 50, 30, 10, 0, and 10 below zero. Color the cold temperatures in dark blue and the cool temperatures in light blue. Use pink for the warm temperatures and red for the hot temperatures.
3. What is a Celsius thermometer? How is it different from a Fahrenheit thermometer?

   Write down some Fahrenheit temperatures on a piece of paper. Find out what those temperatures are on a Celsius thermometer.
4. What's the temperature inside your refrigerator? Inside the freezer? Put a thermometer first in the refrigerator, then in the freezer. Report to the class.
5. Get a newspaper weather report that shows the temperatures around the country. Which city has the hottest temperature? Which has the coolest temperature? What is the temperature in your area?
6. As you work through this section, you may want to find out more about weather. You can find out more by looking in an encyclopedia or by getting books about weather from a library. You can also talk to an expert, such as a scientist or a TV weather reporter.

   Here are some things you may want to find out:
   - What is the atmosphere? What is it made of? How does it help make weather?
   - How does sunlight help make weather? What kind of light rays reach Earth? What is the greenhouse effect?
   - What is climate? How is climate different from weather?

Unit 2

How Windy Is It?

Suppose you heard a weather reporter say:
"It's cool and breezy today."

You know the reporter is describing the amount of heat in the air. (*Cool* tells you that.) But the reporter is also describing something else about the weather—**wind**. (*Breezy* means a little windy.)

Winds are part of weather. Just like heat, winds are always changing. And when they change, the weather changes too.

- What is wind?
- What can cause wind?
- How can wind change the weather?

You'll learn the answers in this unit.

Before You Start

You'll be using the science words below. Find out what they mean. Look them up in the Glossary. On a separate piece of paper, write what the words mean.

1. **block**
2. **instrument**
3. **rotate**

Moving Air

Hold a sheet of paper in front of your face. Now move that paper quickly back and forth, like a fan. What do you feel on your face?

Right! You feel air. But the air was moving. Air that moves is called wind. So what you felt was wind, like the wind outdoors.

Air is easy to move. You moved it by fanning the paper. But you moved only a small amount of air. The wind outdoors is a large amount of air—a block. What do you think makes that block of air move?

One answer is heat. When air gets heated, it moves. Then cooler air moves to where that air was. And a wind blows.

When air is heated, it moves a certain way. How does it move? Do the experiment and find out.

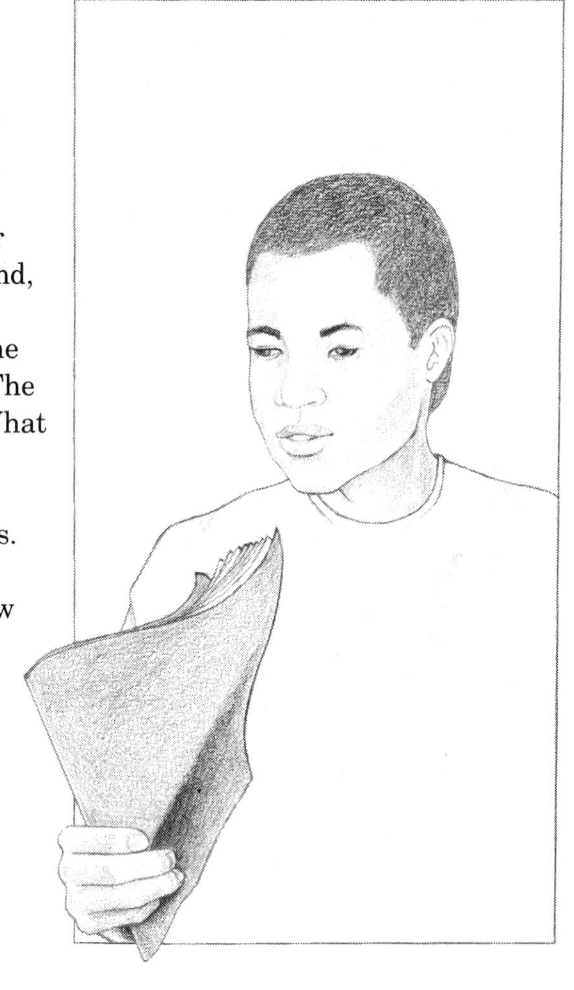

Experiment 1

How does air move when it is heated?

One pin (such as a pushpin or thumbtack)

Materials (What you need)

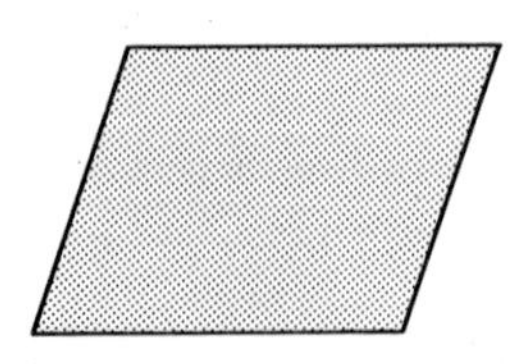

One square piece of foil (about 5 inches square)

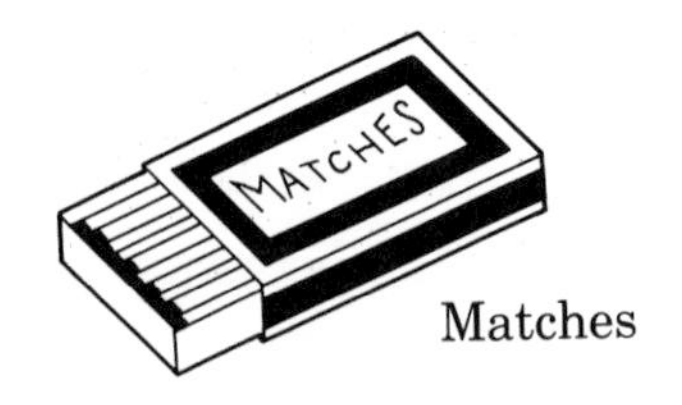

Matches

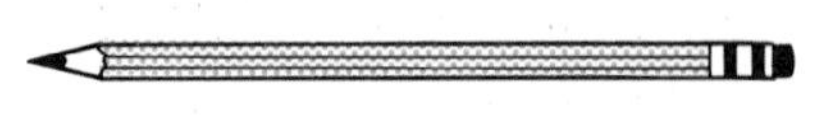

One long pencil with an eraser

One short candle

Procedure (What you do)

1. Push the pin into the center of the foil. Then push the pin into the eraser of the pencil.
2. Light the candle. It will heat the air above it. Hold the candle below the foil. Wait two minutes. What happens to the foil?

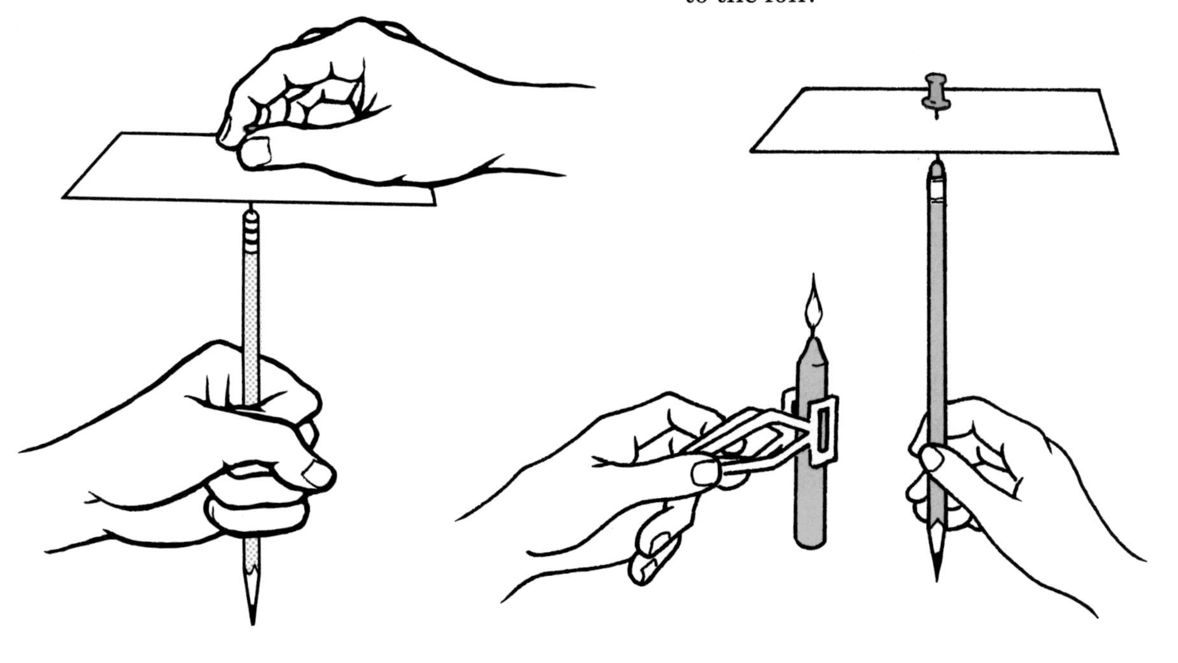

Observations (What you see)

What happens to the foil when you put the candle under it?

Right! The foil moves. The heated air pushes up the foil.

Conclusions (What you learn)

1. What makes the foil move?
   a. cold air above the candle
   b. heated air above the candle
2. How does heated air move?
   a. It moves up.
   b. It moves down.

Experiment 2

What happens when heated air rises?

When you did Experiment 1 on pages 104–105, you saw that air rises when it is heated. The air outside also rises when it is heated. The sun heats the things on Earth, such as the ground and the oceans. Those things heat the air above them. And the heated air rises.

When heated air rises, something happens to the colder air around it. What happens? Do the experiment and find out.

You'll need a hot or cold room that has a door or window that can open. You'll also need a place right outside that room, such as the outdoors, a hall, or another room. That place should be colder or hotter than your room.

Materials

One thin strip of paper (about 6 inches long)

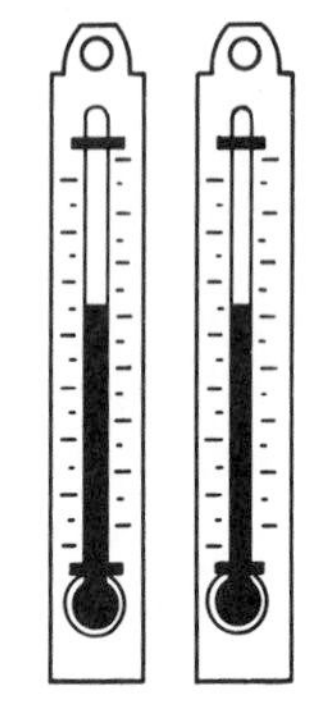

Two thermometers

Procedure

1. Hang a thermometer in the place outside your room. Make sure the thermometer is not in the sun.

2. Hang the other thermometer inside your room. Make sure the thermometer is not in the sun.

   Now close all the doors and windows in the room.

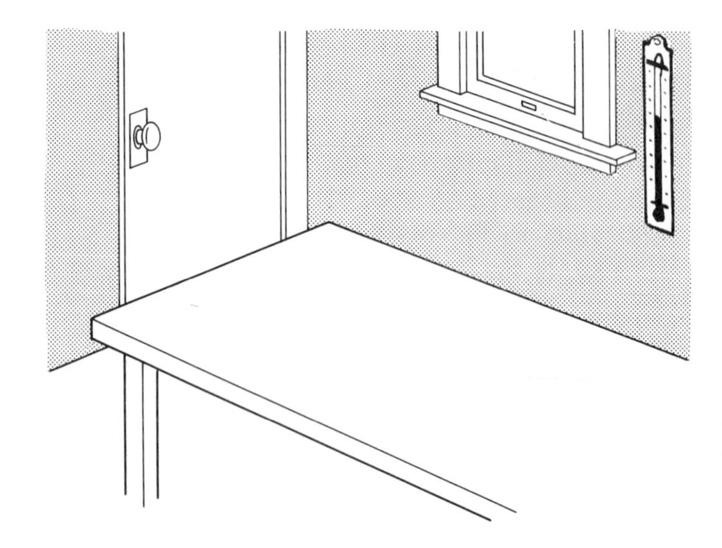

3. Wait ten minutes. Then open the bottom part of a window. (Or open a door.) Hold the strip of paper in front of the open window. (Or hold it above the doorsill.) Look at the strip. Which way does it move: toward the inside of the room or toward the outside of the room?

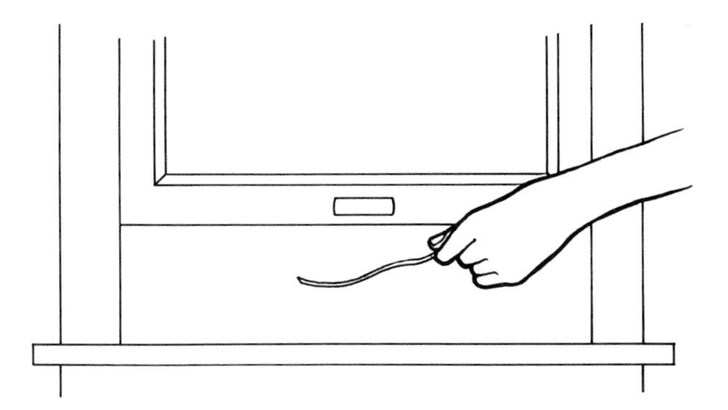

4. Look at the thermometers inside and outside the room. What temperature does each one show?

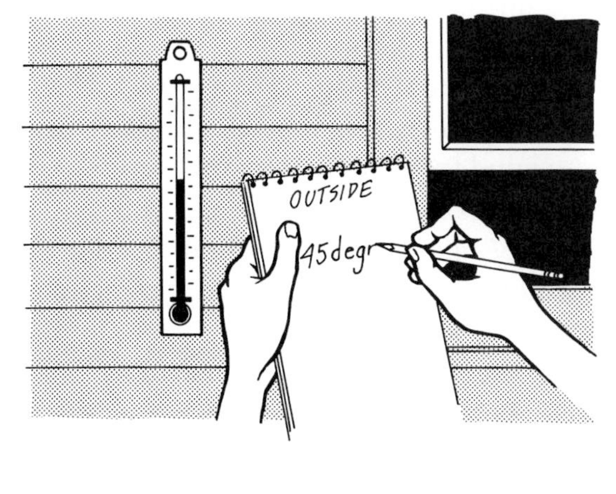

Observations

1. Where is the air hotter?
   a. Inside the room
   b. Outside the room
2. Where is the air colder?
   a. Inside the room
   b. Outside the room
3. What does the strip move toward?
   a. Hotter air
   b. Colder air

Conclusions

1. When you open the window (or door), air moves into or out of the room. That moving air moves the strip of paper.
   a. Where does the moving air come from: hotter air or colder air?
   b. Where does the air move: toward hotter air or toward colder air?
2. You know that air rises when it is heated. What happens to the colder air around it?

   Right! Colder air moves toward the heated air. The colder air rushes to the place where the heated air was. That rushing colder air is wind.

Wind Speed

Winds can be gentle. And they can be very strong. Whether they are gentle or strong depends on how fast they move—their speed.

We measure the speed of a wind by how many miles it travels in one hour. We say that a wind travels so many **miles per hour**. For example, a gentle wind may have a speed of 15 miles per hour. And a wind during a storm may have speeds up to 60 miles per hour.

Weather watchers use a special instrument called an **anemometer**. The anemometer measures the speed of winds. You and your class can make a simple anemometer. Just follow the directions on the next page.

After you've made the anemometer, go outside and measure the wind speed. Keep a record of what you find out. On a separate piece of paper, write the date, the time, and the wind speed.

How to Use Your Anemometer

1. Take a watch and your anemometer to a windy place. Set up the anemometer.
2. Count the number of times the black cup spins around in 30 seconds.
3. Divide that number by 5. Your answer will be the speed of the wind. (For example, suppose the black cup spins 45 times in 30 seconds. *45 divided by 5 is 9*. The wind speed is 9 miles per hour.)

How to Make Your Own Anemometer

What You Need

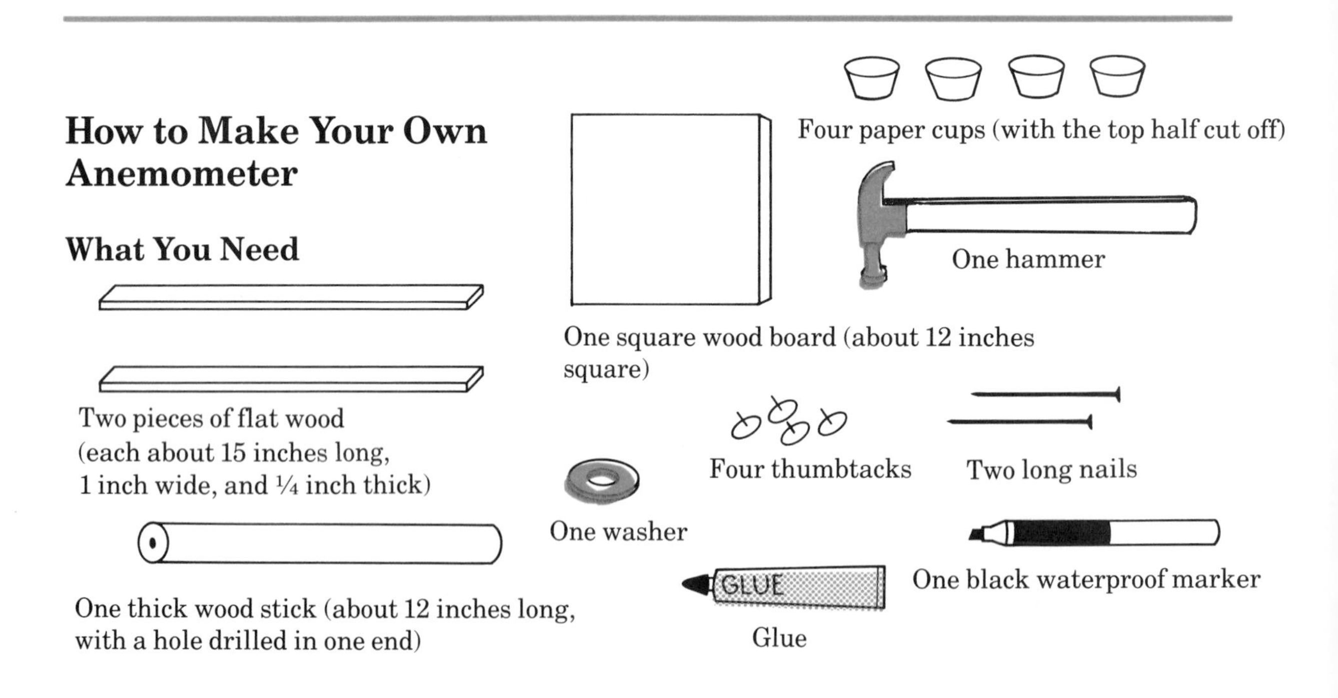

What You Do

1 Put the board on the stick (on the end that doesn't have a hole). Nail the board to the stick.

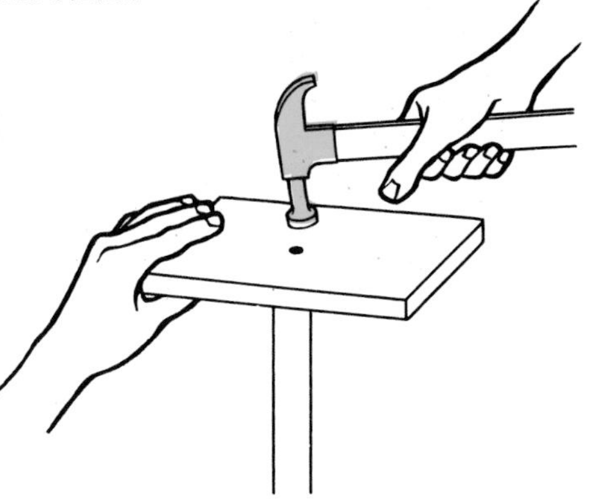

2 Tack one cup onto the side of one end of a piece of flat wood. Turn that piece of wood over. Tack another cup to the side of the other end.

Tack the other cups on the other piece of wood in the same way.

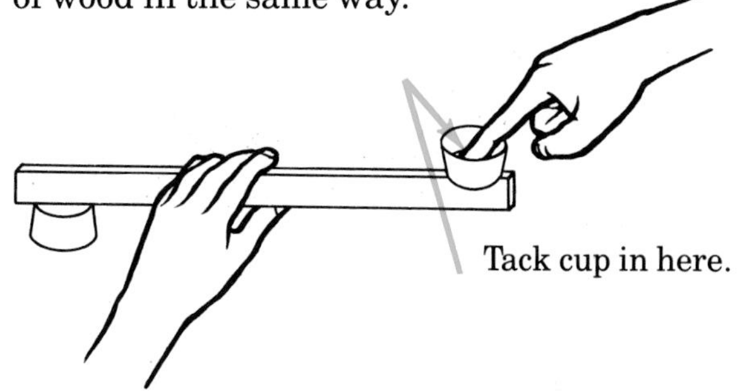

3 Use the black marker to color one of the cups. Color it completely.

4 Glue the two pieces of wood together in the form of a cross.

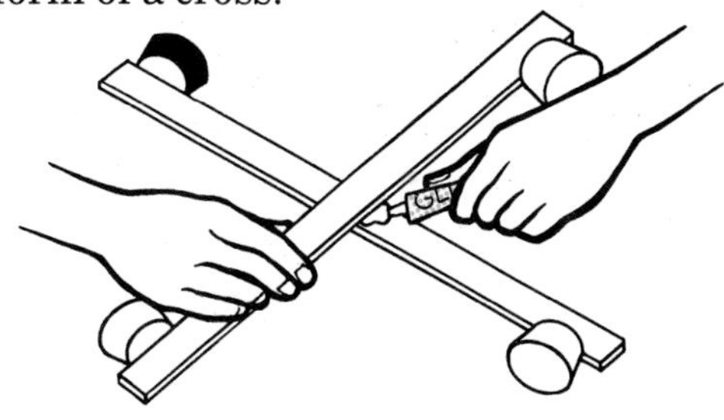

5 Hammer a nail into the center of the cross. (Hammer it on top of a thick stack of newspapers.)

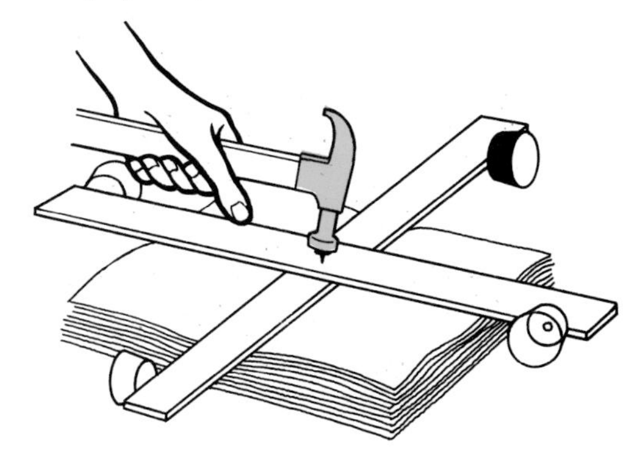

6 Put the washer over the hole in the stick. Fit the nail into the washer and the hole. The anemometer is now ready to use.

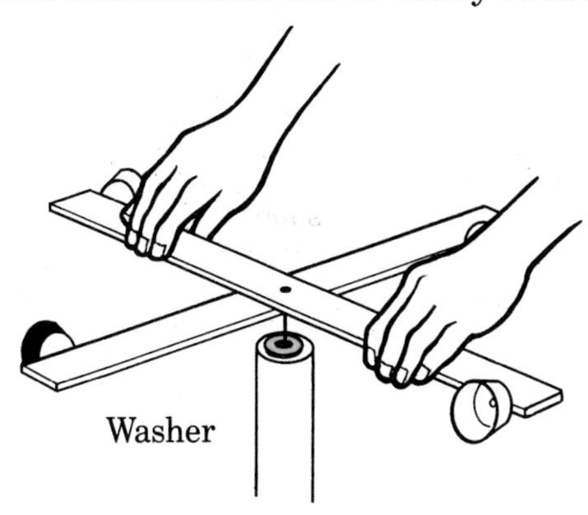

A tornado

Winds Change Weather

When you did the two experiments, you saw how small amounts of air move. Large blocks of air above Earth move in the same way. They become heated and rise. And cooler air—winds—rush in to take their place.

But heat is just one thing that causes wind. The way Earth moves also causes wind. How does Earth move?

Right! Earth spins, or rotates. Its rotation makes air move.

As the air above the land and oceans becomes heated, it rises. Then it cools off and sinks down again. Earth's rotation moves that sinking air in different directions. That moving air becomes wind.

When weather reporters describe winds, they tell what direction the winds come from. For example, a wind from the north is called a north wind.

Winds from different directions have different temperatures. And when they meet air that's warmer or colder, they change the weather. They bring snow or rain. Or they stop a storm.

Winds can also cause storms. One example of a windstorm is a **tornado**. In a tornado, winds whirl round and round at terrific speeds and with terrific force. What other kinds of windstorms do you know about?

Weather Watch

You've learned how to measure wind speed with an anemometer. Here's a way to guess wind speed without an anemometer.

You'll need trees and the wind chart on this page. The chart shows what the wind speed might be when different parts of trees move. For example, if small branches move, the wind is about 13 to 18 miles per hour.

On a separate piece of paper, write today's weather report. List these things:

1. Date
2. Time
3. Temperature
4. Wind speed

Use a thermometer to find the temperature, and an anemometer or the wind chart to find the wind speed.

Wind Chart

How the Trees Move	Wind Speed (miles per hour)
Leaves move just a little.	4–7
Leaves and small twigs move.	8–12
Small branches move.	13–18
Small, leafy trees sway.	19–24
Large branches move.	25–31
Large trees sway.	32–38

Review

Show what you learned in this unit. Finish the sentences. Match the words on the left with the correct words on the right.

1. When heated air rises, cooler air
2. When heated air cools,
3. Winds move in different directions
4. Wind is the cooler air that rushes to the
5. Weather can change when
6. Wind is measured by

a. moves to where that air was.
b. place where heated air was.
c. it sinks back to Earth.
d. because Earth rotates.
e. how many miles per hour it travels.
f. winds meet air that's warmer or colder.

Check These Out

1. Find out how a weather vane works. On a separate piece of paper, draw a picture of one. Explain how it works to the class. Or find out how to make one, and then make it.
2. For one week, keep a daily record of the wind direction.
3. Suppose you were living on the beach. You would have a sea breeze during the day. And you would have a land breeze during the night. Find out why that is true.
4. There are many kinds of windstorms. Find out about one. Make a drawing or a story about what it might be like if such a windstorm hit your area.
5. Find out what the words *wind chill factor* mean. Then explain this sentence to the class: *"It's 25 degrees with a wind chill factor of 16 below zero."*
6. Here are more things you may want to find out:
   - How does the atmosphere circulate? How would it move if Earth didn't rotate?
   - What's a cold front? A warm front? What kind of weather does each bring?
   - What are prevailing winds? Trade winds? Horse latitudes? Doldrums? How did they get their names?
   - What kind of weather does an east wind bring? A west wind? A north wind? A south wind?

Unit 3

Heavy Air

You learned that weather reporters describe weather by talking about temperatures and winds. They also describe weather by talking about the weight of air.

Air can be heavy, or air can be light. But its weight is always changing. When the weight of air changes, weather also changes.

- How do weather reporters describe the weight of air?
- How does weather change when the weight of air changes?
- How is the weight of air measured?

You'll learn the answers in this unit.

Before You Start

You'll be using the science words below. Find out what they mean. Look them up in the Glossary. On a separate piece of paper, write what the words mean.

1. **air mass**
2. **air pressure**

KGO-TV, Channel 7, San Francisco

A weather reporter describes air pressure.

Heavy and Light

Imagine this:

It's a cold and windy day. You turn on a TV weather report. The reporter says:

"Low pressure air is over us."

What do you think the reporter is talking about?

The reporter is talking about the weight of air pushing on Earth.

When reporters describe the weight of air, they don't tell us how much air weighs. Instead, they tell us how hard the air presses against Earth—its pressure. They tell us if that pressure is *high* or *low*.

Heavy air presses harder than light air. So heavy air has a high pressure.

Light air doesn't press as hard. Light air has a low pressure.

Air pressure is always changing. It becomes higher or lower. When it starts getting higher, weather reporters say the air pressure is *rising*. When it starts getting lower, reporters say the air pressure is *falling*.

Is air getting heavier or lighter when its pressure rises?

Is air getting heavier or lighter when its pressure falls?

Right! Air is getting heavier when its pressure rises (gets higher). And air is getting lighter when its pressure falls (gets lower).

National Weather Service Forecast

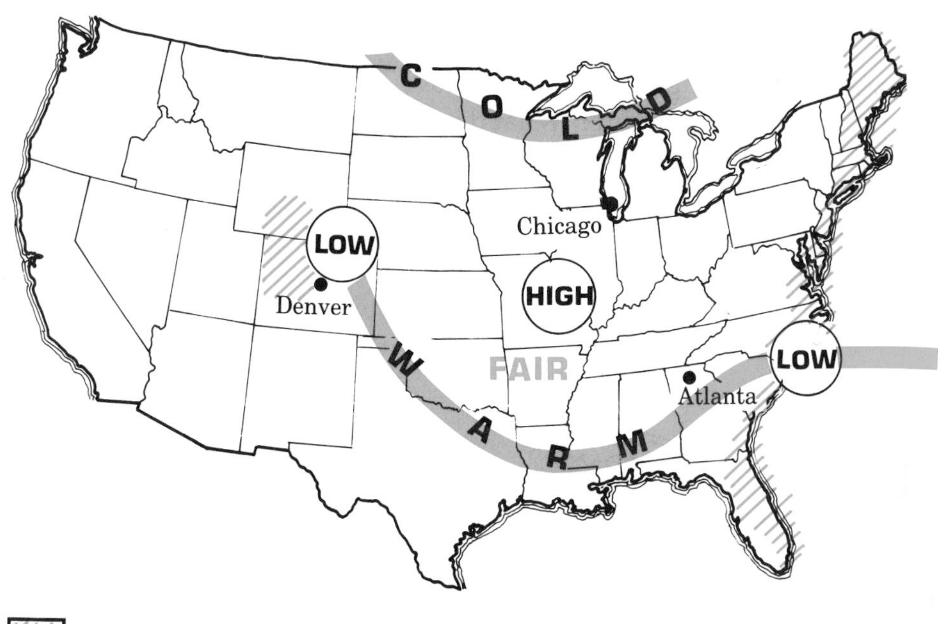

This stands for rain.

High and Low

You learned that air moves. Sometimes an area of moving air is hundreds of miles wide. That large area of air is called an air mass.

Air masses move far over our land. Some air masses have high pressures. And some have low pressures.

Different kinds of air masses can change the weather. For example, an air mass with a high pressure may bring fair weather. But an air mass with a low pressure may bring rain or snow.

Look at the weather map on this page. That map is from a newspaper weather report. The map shows the air pressure over three cities. And it shows what the weather will be like in those cities.

What is the air pressure over Denver?

Right! The air pressure is low.

What will the weather be like in Denver?

Right! It's going to rain in Denver.

What is the air pressure over Atlanta?

What will the weather be like in Atlanta?

What is the air pressure over Chicago?

What will the weather be like in Chicago?

Experiment 3

Which is heavier: warm air or cold air?

Air pressure is always changing. So the air above you is always getting heavier or lighter. What do you think makes air heavier or lighter?

One answer is the temperature of the air—how warm or cold the air is. Which is heavier: warm air or cold air? Do the experiment and find out.

Two pieces of cotton string (each 1 foot long)

Materials

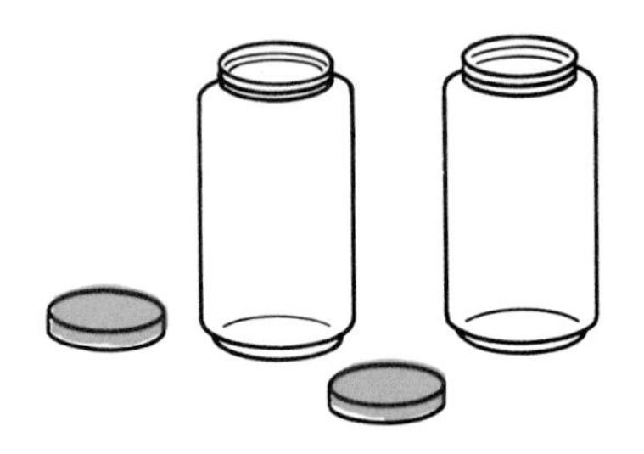

Two empty pint jars and their covers

One bowl of ice and water

One bowl of warm water

Matches

Procedure

1. Roll or twist each piece of string into a loose ball. Set the strings on fire. Drop them into the jars.

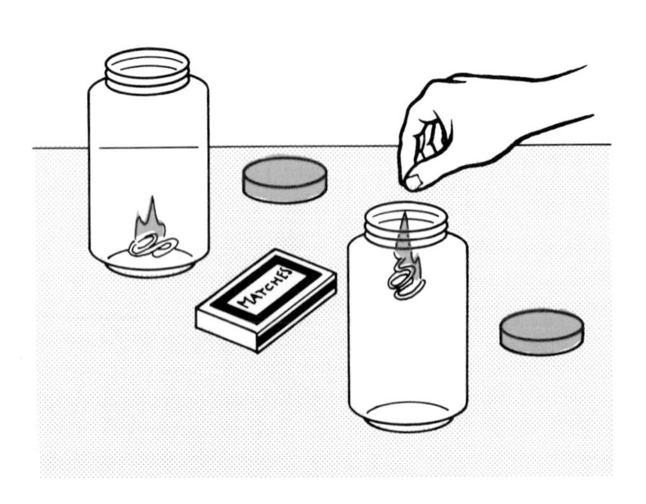

2. Cover both jars tightly.

3. Put one jar into the bowl of warm water. Put the other jar into the ice water. Let the jars fill up with smoke.
4. When the jars are filled with smoke, take them out of the bowls. Open both jars. Quickly turn them upside down.

Observations

When you open the jars, smoke rises out of them. The smoke lets you see what happens to the air in the jars. The smoky air that rises faster is lighter. Which rises faster: the heated air or the cooled air?

Conclusions

1. Which is heavier: warm air or cold air?
   Right! Cold air is heavier.
2. What kind of air has a low pressure: hot or cold?
3. What kind of air has a high pressure: hot or cold?

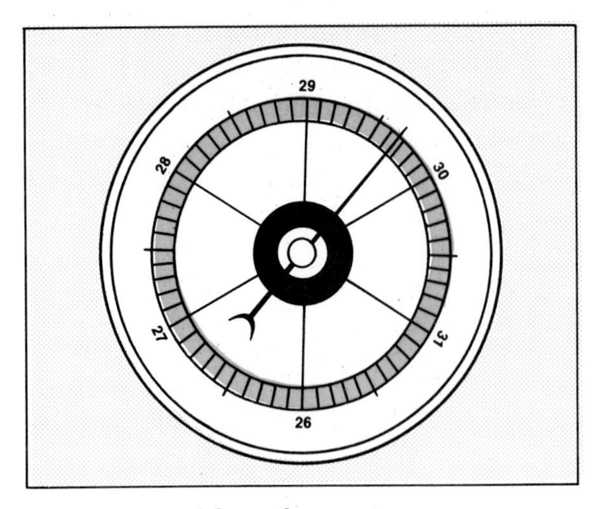

A home barometer

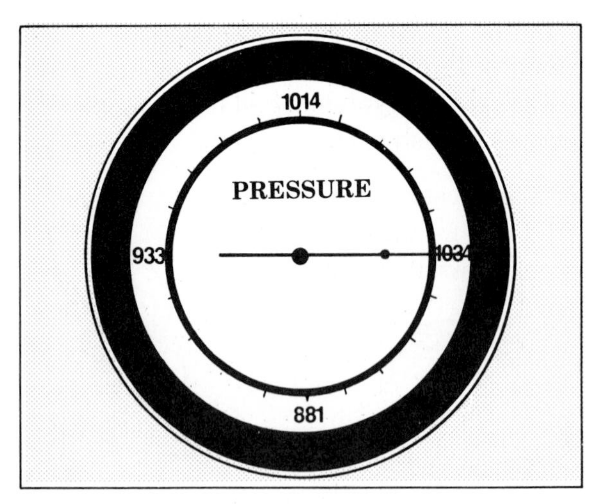

A barometer that's used in weather stations

Weather Watch

A **barometer** is an instrument that measures the pressure of air. The pictures on this page show two kinds of barometers.

Many people keep barometers in their homes. If you know someone who has a barometer, ask to borrow it and bring it to class.

Then, on a separate piece of paper, write a weather report for today. List these things:

1. Date
2. Time
3. Temperature
4. Wind speed
5. Air pressure

If you don't have a barometer, you can find out the air pressure in one of these ways:

- Telephone a TV station and speak to a weather reporter.
- Watch a TV weather report for your area.
- Get a weather map, like the one on page 115, from a newspaper.
- Telephone a National Weather Service office in your area.

Review

Show what you learned in this unit. Find the right words to finish each sentence.

1. Air pressure is
   a. how fast air moves.
   b. how hard air presses against Earth.
2. When air has a high pressure
   a. the air is heavy.
   b. the air is light.
3. When air pressure rises
   a. the air is getting heavier.
   b. the air is getting lighter.
4. A low air pressure means
   a. fair weather.
   b. rain or snow may fall.
5. Cold air is
   a. heavier than warm air.
   b. lighter than warm air.
6. Air pressure is measured by
   a. a barometer.
   b. an anemometer.

Check These Out

1. Air pressure is always changing. When air pressure changes, the weather also changes. For three days, keep a record of the air pressure and the weather in your area. Find out what the air pressure is in the morning and in the afternoon. Then describe the weather at those times (fair, windy, and so on).
2. Find out how a mercury barometer works. What happens to a mercury barometer when air pressure falls? When air pressure rises?
3. In some towns and cities, people can dial a phone number for a weather report. Find out if your town or city has such a number. Look in the phone book for *weather*. Call that number.
4. Here are more things you may want to find out:
   - What are molecules? What happens to molecules in the air when the pressure changes?
   - Who was Evangelista Torricelli? What did he discover?
   - What are cyclones? What are anticyclones? How are they different from each other? How do winds circulate in cyclones and anticyclones?

Unit 4

Water in the Air

Did you know this? The air around you has water in it. That water helps make weather.

Air can have a lot of water in it. Or it can have a little. The amount of water in the air is always changing. When the amount changes, weather sometimes changes too. So weather reporters often tell us how much water is in the air.

- How does water get into the air?
- Why does the amount of water in the air change?
- How do weather reporters talk about water in the air?

You'll learn the answers in this unit.

Before You Start

You'll be using the science words below. Find out what they mean. Look them up in the Glossary. On a separate piece of paper, write what the words mean.

1. **condense**
2. **evaporate**
3. **humid**

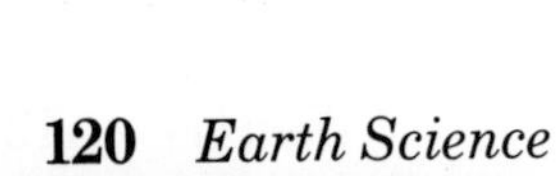

Water You Can't See

Let's say it's raining. The rain makes puddles of water on the sidewalk. Then it stops raining and the sun comes out. What happens to the water in the puddles?

Right! The water dries up. It *evaporates*.

Where do you think the water goes?

Right! Water goes into the air.

Water can be in three forms: It can be a solid, a liquid, or a gas. When water freezes and turns into ice, it is a solid. When solid water (ice) melts, it turns into liquid water. And when liquid water evaporates, it turns into a gas that goes into the air. We call that gas **water vapor**.

Liquid water is evaporating all the time. It evaporates from oceans and lakes, from rivers and pools. It evaporates from trees and other plants. It evaporates from anything that is wet.

As liquid water evaporates, water vapor floats into the air. When the water vapor stays in the air, we say the air "holds water."

What do you think causes liquid water to evaporate?

The answer is heat.

You've seen many times how heat causes water to evaporate. For example, you've seen how wet clothes become dry when they are in the sun.

What's another example of how heat causes water to evaporate?

How Much Water Vapor?

It's a hot, uncomfortable day. The air feels heavy and sticky. You listen to a weather report. The reporter says:

"The humidity is high today."

What do you think the reporter is talking about?

The reporter is talking about how much water vapor is in the air—the **humidity**.

Humidity is the amount of water vapor the air in one place holds. A *high humidity* means that the air holds a lot of water. What does a *low humidity* mean?

Right! A low humidity means that the air holds little water.

Air that has a low humidity can take in much more water vapor. So liquid water evaporates quickly. More and more water vapor goes into the air. And wet things dry quickly.

As the air fills with water vapor, the air's humidity rises. Now the air can't take in as much water vapor. So liquid water evaporates slowly. And less and less water vapor goes into the air. Do wet things dry quickly or slowly when the humidity is high?

When the humidity is high, we say the weather (or air) is humid. In humid weather, our skin feels wet and sticky. Why do you think this happens?

Right! Our sweat can't evaporate easily in humid weather. So we feel wet and sticky.

Back into a Liquid

Have you ever seen this? Tiny drops of water on grass, leaves, and cars early in the morning.

Suppose that water hasn't come from rain. Where do you think it comes from?

Right! That water comes from the air.

You learned that liquid water evaporates. It changes into water vapor. And the water vapor goes into the air.

But the water vapor in the air can change too. It can change back into liquid water. When that happens, we say the water vapor *condenses*.

When large amounts of water vapor condense, the weather changes. For example, water vapor can condense and make **fog**. In a fog, millions of tiny drops of liquid water float in the air close to the ground.

You can easily make water vapor condense. Do this: Breathe hard on a cold glass window. What happens to the place you breathe on?

The place you breathe on is covered with tiny drops of water. Water vapor in your warm breath condenses when it touches cold glass.

Now do this: Fill a glass with ice and cold water. Put that glass in a warm place. What happens to the outside of the glass? Why does it happen?

Experiment 4

Which holds more water vapor: warm air or cold air?

You learned that air masses are always moving over the land. They carry water vapor with them. And they can change the weather in the places they go to.

One air mass may hold more water vapor than another air mass. How much water vapor a mass holds depends on the temperature of its air.

Does warm air or cold air hold more water vapor? Do this experiment and find out.

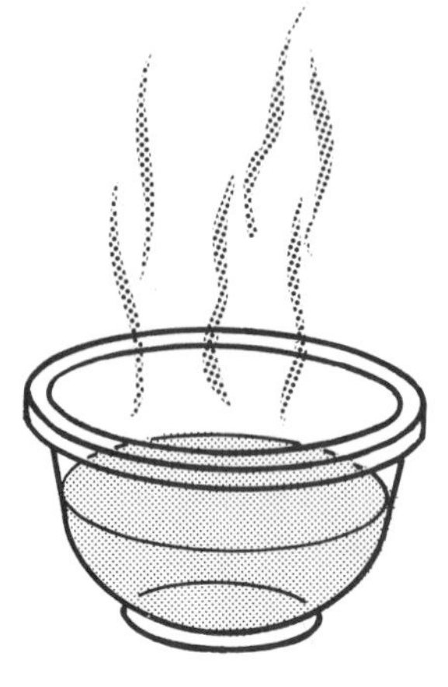

One bowl of warm water

Materials

One small glass jar with a tight lid

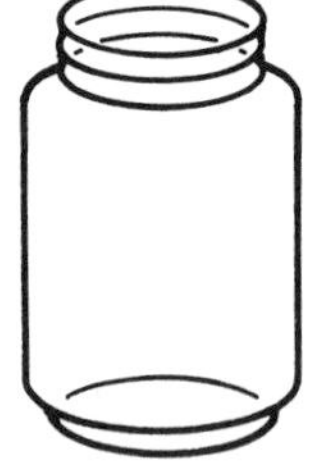

One bowl of ice water

Procedure

1. Put water vapor in the jar this way: Breathe hard into the jar. Stop when the inside is covered with condensed water.

2. Quickly put the lid tightly on the jar.

3. Put the jar into the bowl of warm water. Wait one minute. Then take the jar out. What happens to the condensed water inside the jar?

4. Now put the jar into the bowl of ice water. Wait one minute. Take the jar out. What happens to the glass inside the jar?

Observations

1. What happens when the jar is put into warm water?
   a. More water condenses inside the jar.
   b. Liquid water inside the jar evaporates.
2. What happens when the jar is put into ice water?
   a. Water condenses inside the jar.
   b. Liquid water inside the jar evaporates.

Conclusions

1. Use the word *more* or *less* to finish each sentence.
   a. When the jar is put into warm water, the air in the jar becomes warm. The liquid water evaporates. It changes into water vapor. The air now holds ________ water vapor.
   b. When the jar is put into ice water, the air in the jar becomes cold. The water vapor in the air condenses. It changes into liquid water. The air now holds ________ water vapor.
2. Which holds more water vapor: warm air or cold air?

   Right! Warm air holds more water vapor than cold air.

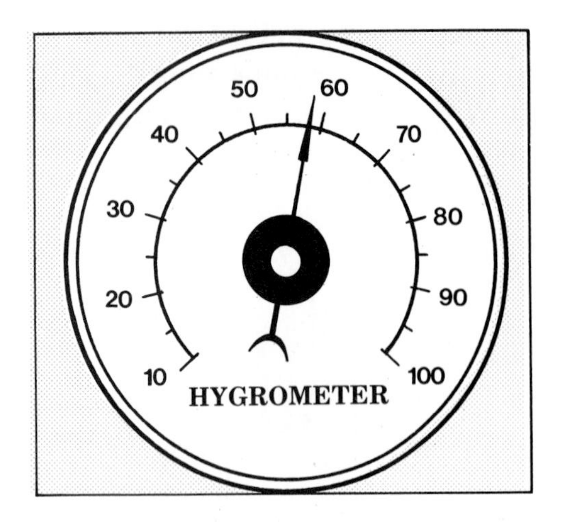

A hygrometer measures humidity.

Weather Watch

Weather people measure humidity with an instrument called a **hygrometer**. You probably don't have a hygrometer. But you can usually guess if the humidity is high or low. How? By looking for certain signs.

The charts on this page list some signs of high and low humidity. Use them to see if the humidity today is high or low. Then, on a separate piece of paper, write a weather report. List these things:

1. Date
2. Time
3. Temperature
4. Wind speed
5. Air pressure
6. Humidity

Signs of High Humidity	Signs of Low Humidity
• The air feels damp and heavy.	• The sky is clear and has no clouds.
• Water condenses quickly on glasses of cold drinks.	• Clouds are getting smaller.
• Sweat stays on the skin, so the skin feels sticky.	• The skin feels dry.
• Wet clothes take a long time to dry.	• Clothes dry quickly.
• Clouds build up in the sky.	• Often the temperature is very low.

Review

Show what you learned in this unit. Finish the sentences. Match the words on the left with the correct words on the right.

1. Heat makes liquid
2. Water vapor is
3. Humidity is the amount
4. Water vapor condenses
5. Fog is made of
6. Weather people use hygrometers to

a. of water vapor in the air.
b. water evaporate.
c. the gas form of water.
d. measure the humidity of the air.
e. tiny drops of condensed water.
f. into liquid water.

Check These Out

1. Look for the signs of high and low humidity listed on page 126. Keep a record of what you see and feel for one week. Tell whether the humidity was high or low for each day.
2. On the West Coast, summers are often cold and foggy near the ocean. Find out why.
3. Here are more things you may want to find out:
   - What happens to the molecules in water when water evaporates? What happens to the molecules when water condenses?
   - Why does heat make water evaporate quickly?
   - What is relative humidity? What is dew point? How are humidity, relative humidity, and dew point connected to each other?
   - How much water does Earth have? Where is most of that water found—on Earth or in the atmosphere?
   - Which parts of the world always have low humidity? Which parts always have high humidity? Why do those places always have low humidity or high humidity?

Unit 5

Water from the Sky

Every day, tons of water evaporate. All that water goes into the air as water vapor. Sooner or later, most of that water vapor condenses.

Water vapor condenses on or near the ground. It also condenses in the sky, high above the ground.

Sooner or later, most of the water in the sky falls to the ground. It falls as rain, snow, sleet, and hail. When it falls, the weather changes.

- How does water condense in the sky?
- How does water fall from the sky?
- What causes snow, sleet, and hail?

You'll learn the answers in this unit.

Before You Start

You'll be using the science words below. Find out what they mean. Look them up in the Glossary. On a separate piece of paper, write what the words mean.

1. **frozen**
2. **precipitation**

Sticking to Something Solid

You learned that water vapor condenses into liquid water. But it won't condense unless it has something to stick to. That thing must be solid.

Suppose you take a hot shower. The air in the bathroom becomes hot. It holds a lot of water vapor. That water vapor condenses when it touches something solid. What does it condense on?

Right! It condenses on solid things, such as walls and mirrors.

We can't always see the things water vapor condenses on. For example, fog is made up of tiny drops of condensed water that float in the air. What do you think water vapor condenses on to make fog?

Air is filled with millions of tiny solids. Some of the solids are dust. Some are tiny bits of ash that come from smoke. Water vapor condenses on those tiny solids. And we see fog.

Water vapor condenses the same way in the sky. It condenses on tiny solids that float high in the air. Sometimes the tiny drops of condensed water gather together. What do you think they form when that happens?

Right! They form **clouds**.

Experiment 5

When does water vapor condense and make clouds?

Water vapor in warm air condenses on tiny solids. And the condensed water makes clouds. But water vapor condenses only when something happens to the air it's in. When does water vapor condense? Do this experiment and find out.

Materials

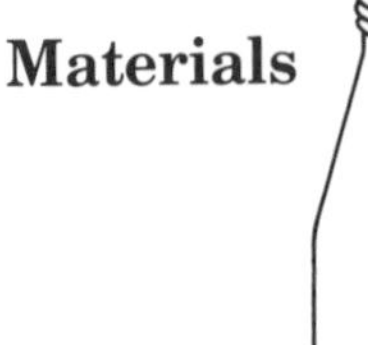

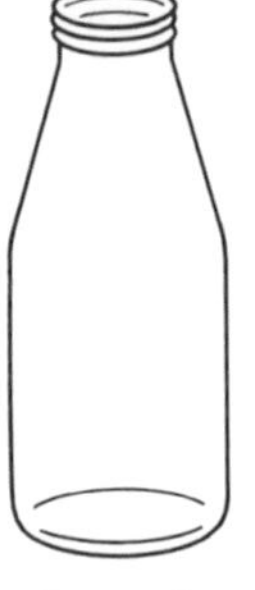

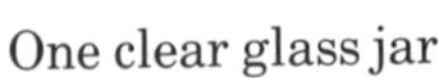

One clear glass jar

Hot water

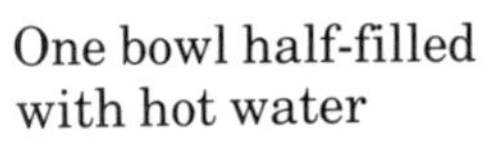

One bowl half-filled with hot water

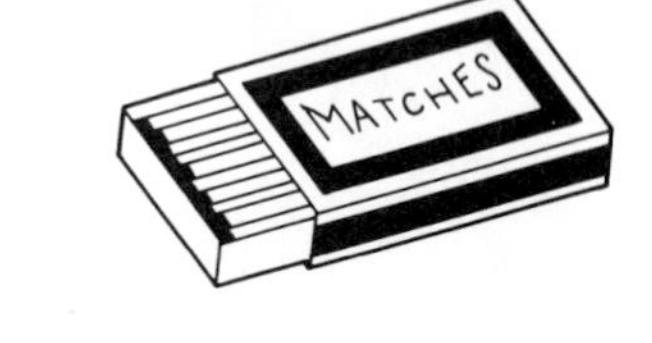

Matches

Ice in a small plastic bag

Procedure

1. First warm the jar. Fill it with hot water. Wait for one minute.

2. Pour the water out of the jar. The air in the jar will become warm. That air will be full of water vapor.

3. Light a match and drop it into the jar. The match will go out right away.

4. Stuff the bag of ice into the top of the jar. Make sure the top of the jar is completely covered. Put the jar into the bowl of hot water.

Observations

1. The match makes something inside the jar. What is it?

   Right! The match makes smoke. Tiny solids in the smoke go into the air inside the jar.
2. What happens inside the jar when you put ice in it?

   Yes! A cloud forms inside the jar. Water vapor condenses on the tiny solids and makes a cloud.

Conclusions

1. What kind of air is in the jar before you put ice on it: warm or cold?
2. What happens to the warm air when you put ice on the jar?

   Right! The warm air near the ice is cooled.
3. Now finish this sentence:

   Water vapor condenses and makes clouds when w ________ air is c ________.

Falling Water

Suppose the tiny drops of water in a cloud grow big and heavy. The cloud can't hold them. What happens to the water?

Right! The water falls to the ground. It falls as rain, sleet, or hail.

Water that falls from clouds is called *precipitation*. Rain, sleet, and hail are different kinds of precipitation.

When a lot of precipitation falls, weather reporters say the precipitation is heavy. What do you think they say when a little precipitation falls?

Right! The precipitation is light.

Precipitation usually starts from a cloud as **rain**. But sometimes the rain passes through very cold air. It freezes and falls to the ground as drops of ice. We call those drops of ice **sleet**.

Sometimes the rain falls through many layers of cold and warm air. Then the rain freezes and turns into balls of ice. We call those balls of ice **hail**.

Another kind of precipitation also falls from clouds. But it is frozen when it leaves the clouds. What is that frozen precipitation called?

Right! That frozen precipitation is **snow**. Snow is different from sleet and hail. Sleet and hail are frozen water. Snow is frozen water vapor.

Round and Round

Water is always moving. It moves round and round, from the ground to the air, from the air to the ground. As it moves, it changes its form. You've learned the different forms water changes into.

The picture on this page shows how water is always moving. Finish the sentences on the picture. Use the words below. Start with **1**.

clouds	condenses	evaporates
liquid water	precipitation	water vapor

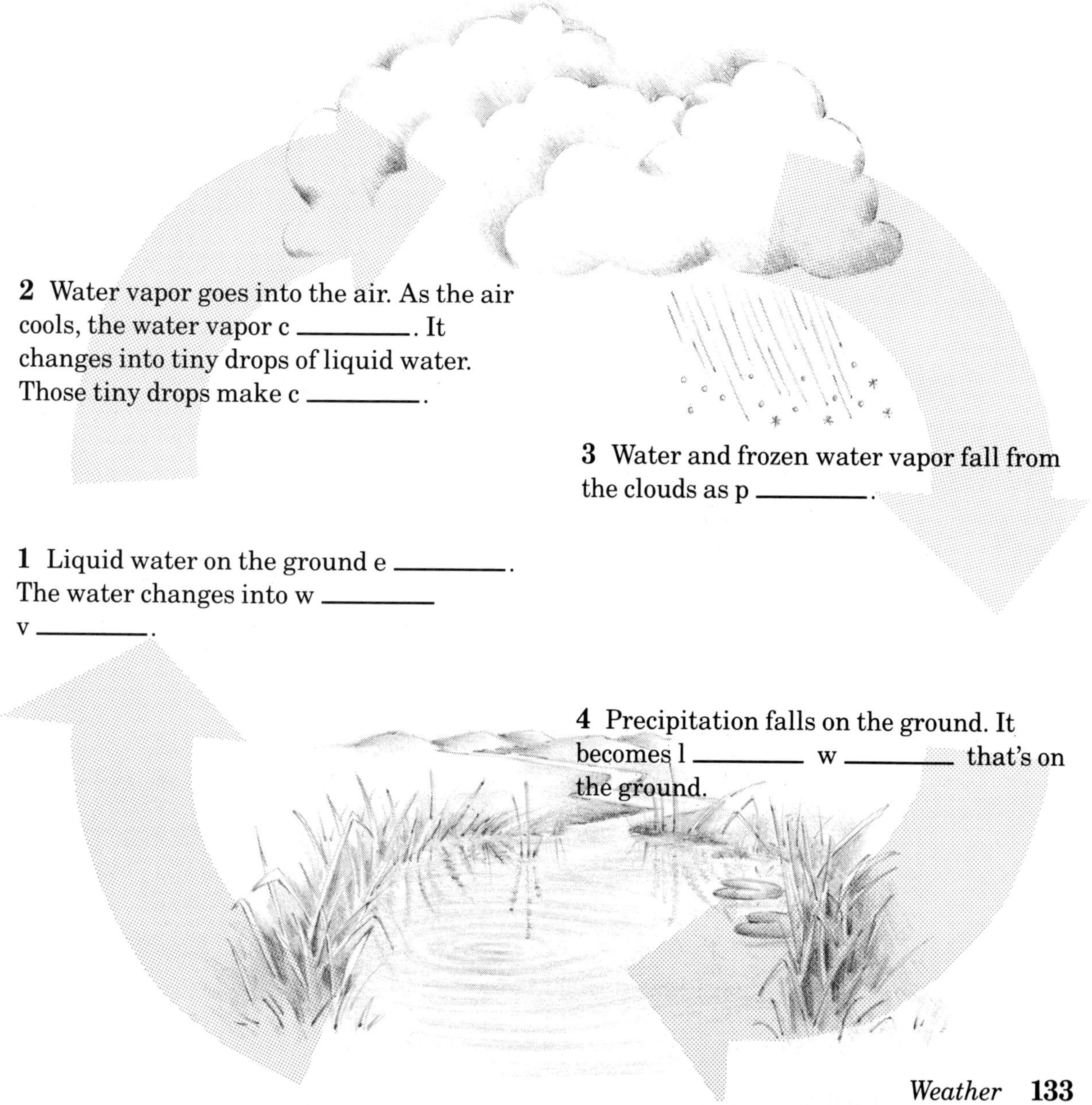

2 Water vapor goes into the air. As the air cools, the water vapor c ________. It changes into tiny drops of liquid water. Those tiny drops make c ________.

3 Water and frozen water vapor fall from the clouds as p ________.

1 Liquid water on the ground e ________. The water changes into w ________ v ________.

4 Precipitation falls on the ground. It becomes l ________ w ________ that's on the ground.

Clouds in the Sky

Are there clouds in the sky right now? What do they look like? Do you think they may bring rain?

Clouds are an important part of the weather. All precipitation comes from clouds. Also, clouds can give clues about how the weather may change.

Scientists have names for the different kinds of clouds you can see in the sky.

Have you seen thin wispy clouds in the sky? These are called *cirrus* clouds. Cirrus clouds are often a sign of an approaching storm.

Clouds that look flat and spread out across the sky are called *stratus* clouds. Stratus clouds often bring light rain.

White, puffy clouds like cotton balls are called *cumulus* clouds. During summer afternoons, cumulus clouds can become dark. Then they are called *cumulonimbus* clouds. The ending *-nimbus* means "rain cloud." These dark clouds can produce heavy showers and thunder and lightning.

Do you see any of these kinds of clouds in the sky now?

Cirrus clouds

Cumulus clouds

Stratus clouds

Cumulonimbus clouds

KGO-TV, Channel 7, San Francisco

Weather Watch

What's the weather like today? On a separate piece of paper, write a weather report. List these things.

1. Date
2. Time
3. Temperature
4. Wind speed
5. Air pressure
6. Humidity
7. Are there clouds in the sky? If there are clouds, describe what they look like.
8. Is fog in the air?
9. Is precipitation falling? If precipitation is falling, what kind is it?
   - Light rain
   - Heavy rain
   - Sleet
   - Hail
   - Light snow
   - Heavy snow

Review

Show what you learned in this unit. Match the words in the list below with the correct clues.

cloud	precipitation	sleet
hail	rain	snow

1. Precipitation that is frozen water vapor
2. Rain, sleet, snow, and hail are different kinds of this.
3. What condensed water vapor makes in the sky
4. Precipitation that is frozen water drops
5. Precipitation that is liquid water
6. Precipitation that is balls of ice

Check These Out

1. Find out what kinds of clouds there are. Find out what kind of weather each kind of cloud brings.
2. What kinds of clouds bring lightning? How dangerous is lightning? What are lightning storms? Find out what to do and what not to do during lightning storms.
3. Find out all you can about hail. Why do water drops turn into big ice balls? What does the inside of a hailstone look like?
4. Here are more things you may want to find out:
   - How big is a raindrop? How is a raindrop made?
   - What does a snowflake look like? Do all snowflakes look alike? Why does a snowflake change shape as it falls?
   - How can scientists make rain fall from clouds?
   - The words below are used in weather reports. What do they mean?

haze	patchy fog
partly cloudy	overcast
mist	variable clouds

Meteorologists work at weather stations.

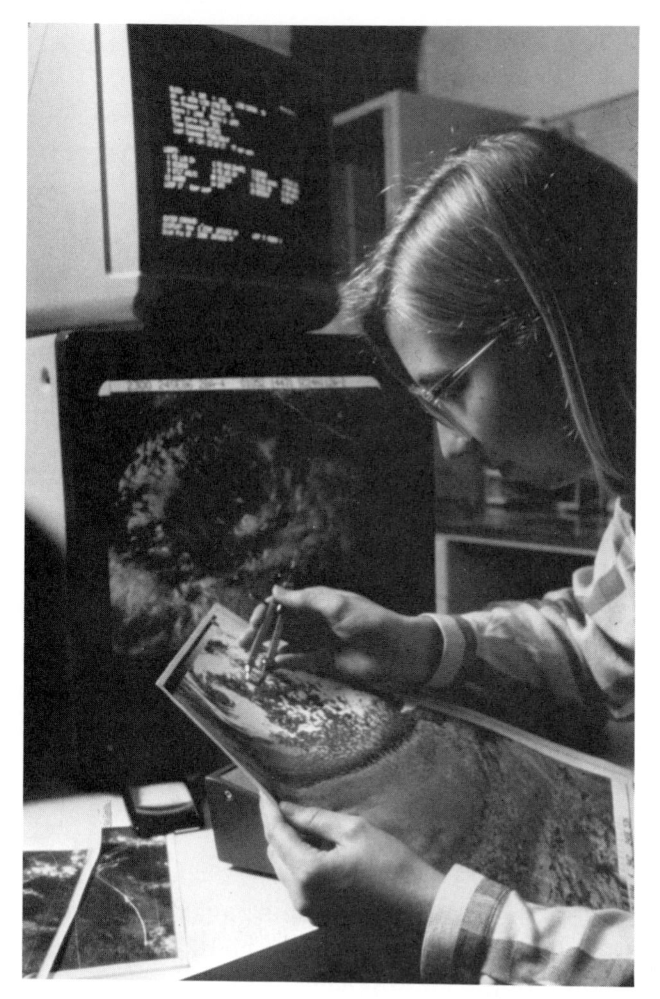

Unit 6

What Will the Weather Be Like?

It's early morning. You're listening to a weather report. The weather reporter says:

"Rain is likely today."

Sure enough, it rains during the day. How do weather reporters know what the weather will be like?

They find out what the weather will be like from **meteorologists**. Meteorologists are scientists who study weather.

Meteorologists work in weather stations. They use special instruments to measure the temperature, pressure, and humidity of the air. They study winds and clouds. They see where air masses are moving. They find out what the weather is like in different places.

Weather moves when air masses move. For example, there may be storms right now in an air mass hundreds of miles away from you. But the weather in your area is fair. If the air mass moves to your area, you'll have storms too.

Meteorologists can tell if the air mass is moving your way. They can **predict** (guess) correctly when the storms will be in your area.

Some weather reports show what meteorologists predict. What do you think those reports are called?

Weather reports that predict weather are called **forecasts**.

Weather Forecast for Today and Tomorrow

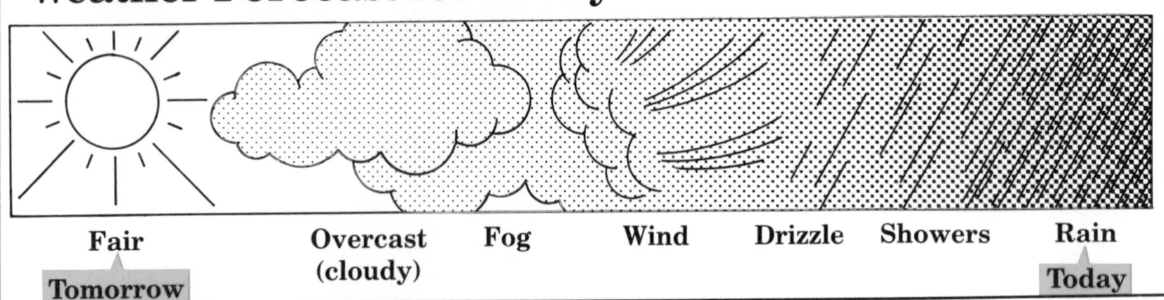

Today and Tomorrow

Rain all day today. Rain ending late tonight. Mostly fair tomorrow. Lows mostly 40s tomorrow night. Highs 60s. Southeasterly winds 15 to 30 mph.

Tuesday Through Thursday

Partly cloudy Tuesday. Fair Wednesday and Thursday. Fog at nights and in the mornings. Highs 50s to low 60s. Lows 30s.

A Weather Forecast

The weather report on this page shows what a forecast in a newspaper might be like. The report is from a Sunday newspaper.

Read the forecast. Then answer these questions:

1. Will it rain today?
2. Will it rain tomorrow?
3. When will the rain end?
4. What will the high temperature be tomorrow?
5. How strong will the winds be tomorrow?
6. What will the weather be like on Wednesday and Thursday?

Weather Watch

Listen to a TV weather report for tomorrow's weather. Or get a weather forecast from a newspaper. Then, on a separate piece of paper, answer the questions about tomorrow's forecast.

1. What will the highest temperature be?
2. What will the lowest temperature be?
3. Will there be any wind?
4. Where will the wind come from?
5. How strong will it be?
6. What will the water in the air be like?
   - Clear
   - Partly cloudy
   - Overcast (cloudy)
   - Foggy
7. Will it rain or snow?
8. What else does the weather forecast predict?

Show What You Learned

What's the Answer?

Choose the correct endings for these sentences. Each sentence has three correct endings.

1. Weather changes when
   a. heat in the air changes.
   b. different kinds of air masses meet.
   c. the amount of water in the air changes.
   d. air pushes down on anything it touches.
2. Air pressure
   a. is always changing.
   b. is how fast air moves.
   c. rises when air gets heavier.
   d. is high when the weather is fair.
3. Water changes in these ways:
   a. Water changes into ice when it melts.
   b. Water changes into water vapor when it evaporates.
   c. Water vapor condenses and forms clouds.
   d. Water changes to ice when it passes through cold air.
4. Some kinds of precipitation are
   a. snow.
   b. water vapor.
   c. rain.
   d. hail.
5. Weather reports can tell you
   a. how high and low the temperatures are.
   b. how strong the winds will be tomorrow.
   c. what time it is.
   d. if it will rain tomorrow.

What's the Word?

Give the correct word or words for each meaning.

1. Water that falls from the sky
   P ________
2. Moving air that is hundreds of miles wide
   A ________ M ________
3. How hard air pushes down on anything it touches
   A ________ P ________
4. Liquid water that changes into a gas
   W ________ V ________
5. The amount of water vapor in the air
   H ________
6. Air that moves quickly
   W ________

Congratulations!

You've learned a lot about weather. You've learned

- What happens when weather changes
- Some ways to predict weather
- How to use weather reports
- And many other important facts about weather

EARTH RESOURCES

What are Earth resources? Where do we find those resources? How do we use Earth resources to stay alive? How can we keep those resources from running out? In this section, you'll learn many facts about Earth's resources. And you'll learn how those resources play an important part in our lives.

Contents

Introduction

Suppose you went to the moon. Could you live there?

Suppose you visited Venus or Mars. Could you live on those planets?

You couldn't live on any planet except Earth! Not unless you brought from Earth the materials you need to stay alive.

Earth is the only planet we know of that has life. Things can live on Earth because it has the materials that living things need in order to stay alive. We call those *vital* materials *Earth resources*.

Earth is made up of other kinds of materials too. We use those materials to make things we need, such as clothes, houses, tools, medicine, and machines. The materials we use to make things are also Earth resources.

In this section, you'll learn about some important Earth resources that we use. You'll learn how we can use those resources wisely.

Right now you're using some Earth resources. What do you think one is?

Unit 1

Air

You can't see air. You can't smell it, and you can't taste it. But it is something you use all the time.

Air is a **vital** Earth resource. That means living things need it to stay alive. If Earth did not have air, nothing would be alive on the planet.

- What is air?
- How is air a vital resource?
- What problems do we have with air?

You'll find out in this unit.

Before You Start

You'll be using the science words below. Find out what they mean. Look them up in the Glossary that's at the back of this book. On a separate piece of paper, write what the words mean.

1. **atmosphere**
2. **particle**

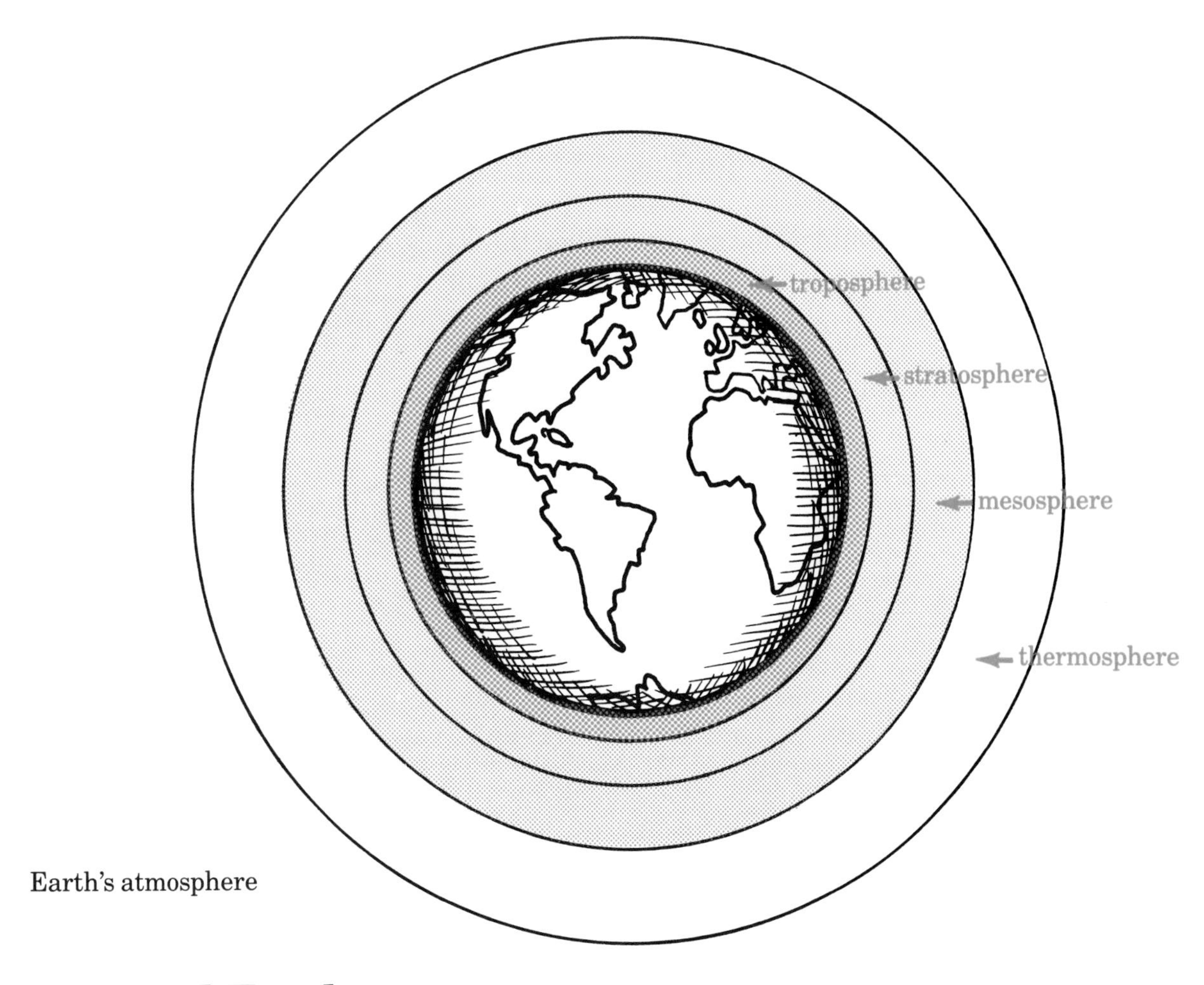

Earth's atmosphere

All Around Earth

Our planet Earth is surrounded by gases. The gases swirl all around Earth. They cover all parts of Earth. Look at the diagram. What do we call the gases that cover Earth?

We call those gases the atmosphere. (Another word for atmosphere is *air*.)

Scientists divide the atmosphere into four layers. You can see the names of the layers on the diagram. Which layer is closest to Earth?

Right! It's the *troposphere*.

The troposphere contains the largest amount of gases. *Nitrogen* is the main gas in the troposphere. Other gases are **oxygen, carbon dioxide,** and **water vapor**.

Earth's plants and animals need certain gases that are in the air. Plants need carbon dioxide. Humans and other animals need oxygen. If they couldn't get those gases, they would die.

Our atmosphere keeps plants and animals alive in another way too: It makes our weather. The air around Earth is always moving. It brings rain clouds to dry places. It moves rain clouds away from wet places. Why is that important to plants and animals?

Experiment 1

Why does a fire need oxygen?

You learned that animals use oxygen. When things burn, they also use oxygen. What happens if a burning thing can't get oxygen?

Materials (What you need)

Two short candles on dishes

Matches

One small glass jar

Procedure (What you do)

1. Light each candle. Wait a minute, until the flames get big.
2. Put the jar over one of the candles.

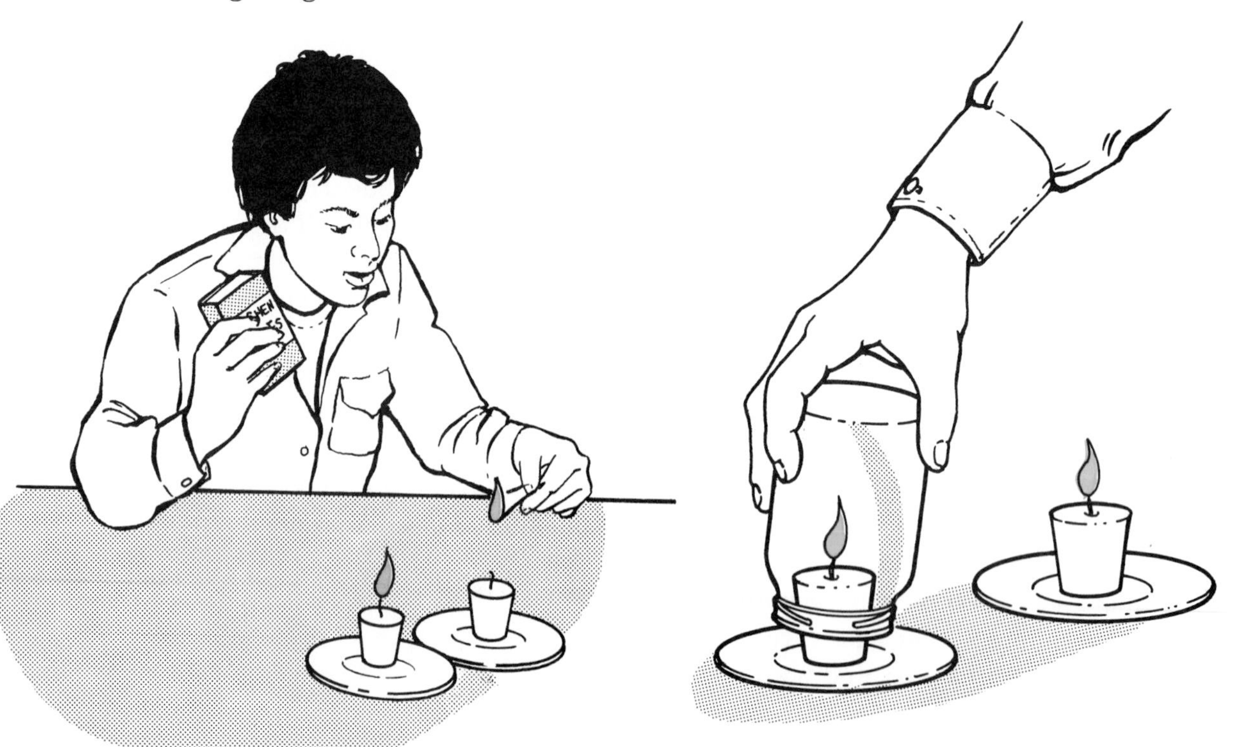

3. Watch both candles. What happens to the flames?

Observations (What you see)

1. What happens to the candle in the jar?
2. What happens to the candle that isn't in the jar?

Conclusions (What you learn)

1. The air outside of the jar had oxygen in it. So did the air inside the jar.
   a. Why did the candle inside the jar go out?
   b. Why did the other candle stay lit?
2. The candle in the jar used up oxygen as it burned. The flame went out slowly. What would happen if you lifted up the jar *before* the flame went out completely? Try it. What happened? Why?
3. Things won't burn unless they have oxygen. If our atmosphere didn't have oxygen, we couldn't make fires. What would happen if we couldn't make fires?

Air Protects Earth

Have you ever been in a greenhouse? That is a kind of glass house that plants grow in. The glass traps heat from the sun. The heat keeps the inside of the greenhouse warm even when the weather is very cold!

Our atmosphere is something like a greenhouse. It keeps Earth warm enough for living things.

Heat energy from the sun must pass through our atmosphere before it reaches Earth. The atmosphere blocks out some of the sun's heat. That keeps Earth from getting too hot.

At night, heat leaves Earth and goes into the air. It starts to go into space. Our atmosphere keeps some heat from going into space. That keeps Earth warm.

Let's suppose we didn't have an atmosphere. What would it be like during the night? During the day?

It would be freezing cold at night. It would be burning hot during the day. It would get too hot and cold for living things.

Our atmosphere protects Earth in one other important way. Objects such as *asteroids* and *meteoroids* travel through space. Sometimes those objects crash into planets. They leave big holes. Earth isn't hit very often by those objects. That's because they usually burn up in our atmosphere. What would Earth look like if it didn't have an atmosphere?

Dirty Air

Imagine it's a summer day. The sun is shining. The sky is bright and blue. But faraway, the air looks brown. It seems dirty. What is wrong with that air?

That air has **pollution** in it. Pollution makes air dirty. In many parts of the world, pollution in air is a serious problem.

Most air pollution comes from things that are burning. Burning makes smoke and harmful gases.

Much of the air pollution we have comes from cars. Cars burn gasoline. This burning makes many kinds of gases. The gases enter the atmosphere.

Many of these gases from cars are harmful by themselves. But they also cause another problem. When sunlight shines on them, they change. They turn into gases that form **smog**.

Smog is bad for living things. It can hurt people's eyes. It can hurt their throats and lungs.

Smog is most dangerous when it builds up over a city. The air turns brown, and people have trouble breathing.

Air pollution also comes from power plants. Power plants burn fossil fuels to make electricity. The smoke from their smokestacks contains certain chemicals. These chemicals go into the atmosphere and pollute it.

The chemicals also combine with water in clouds to make acids. The acids fall to Earth in raindrops. This is called *acid rain.* Acid rain can kill trees and plants. It can kill fish in lakes. How can we stop acid rain?

Earth Watch

Government *agencies* are offices that do special jobs. Some government agencies watch resources such as our air. Those agencies make sure that the resources are safe to use. One such agency is the Environmental Protection Agency. (It is also known as the EPA.)

Find out where an agency like the EPA is in your community. Check the government listings in the phone book. Look for words such as *environmental protection, air quality,* and *pollution control.* On a separate piece of paper, write the name, address, and phone number of the agency.

Now find out how that agency checks your air. Talk with or write to someone there. Ask these questions. Keep a record of what you find out.

1. How do you check the quality of our air?
2. What particles or gases can make our air dangerous?
3. What is being done to keep our air clean?

Review

1. Why is air a vital resource? Choose the three right answers.
   a. It keeps the Earth warm enough for living things.
   b. It gives living things the gases they need to stay alive.
   c. It makes smog.
   d. It carries water to living things.
   e. It takes care of problems in our air.
2. How can our air be harmed? (See page 149.)

Check These Out

1. Make a Science Notebook for this section. Use it to keep a record of what you learn about Earth resources. Put your list of glossary words and their meanings in the notebook. Also keep your notes from experiments and the Earth Watch sections in it. You can keep anything else you learn about Earth resources in your notebook too.
2. Watch a weather report on TV. Find out what weather systems are moving into your community and where those systems are coming from. Then give a weather report to your class.
3. Find pictures in newspapers and magazines of things that cause smog. Cut them out and make a poster.
4. Get a book that has a big picture of Earth's atmosphere. (A librarian can help you find one.) Make a drawing of the atmosphere, and write the name of each layer.

Unit 2

Water

Suppose you just finished a game. You're hot and sweaty. You're thirsty too. You could go for a nice cold drink of water!

Water is another of Earth's resources. It is a vital resource, like air. You drink it because your body must have water. There would be no life on Earth without water!

- How is water a vital resource?
- Where do we get the water we use?
- How do communities make water safe to use?

You'll find out in this unit.

Before You Start

You'll be using the science words below. Find out what they mean. Look them up in the Glossary. On a separate piece of paper, write what the words mean.

1. **evaporate**
2. **ground water**
3. **renewable resource**

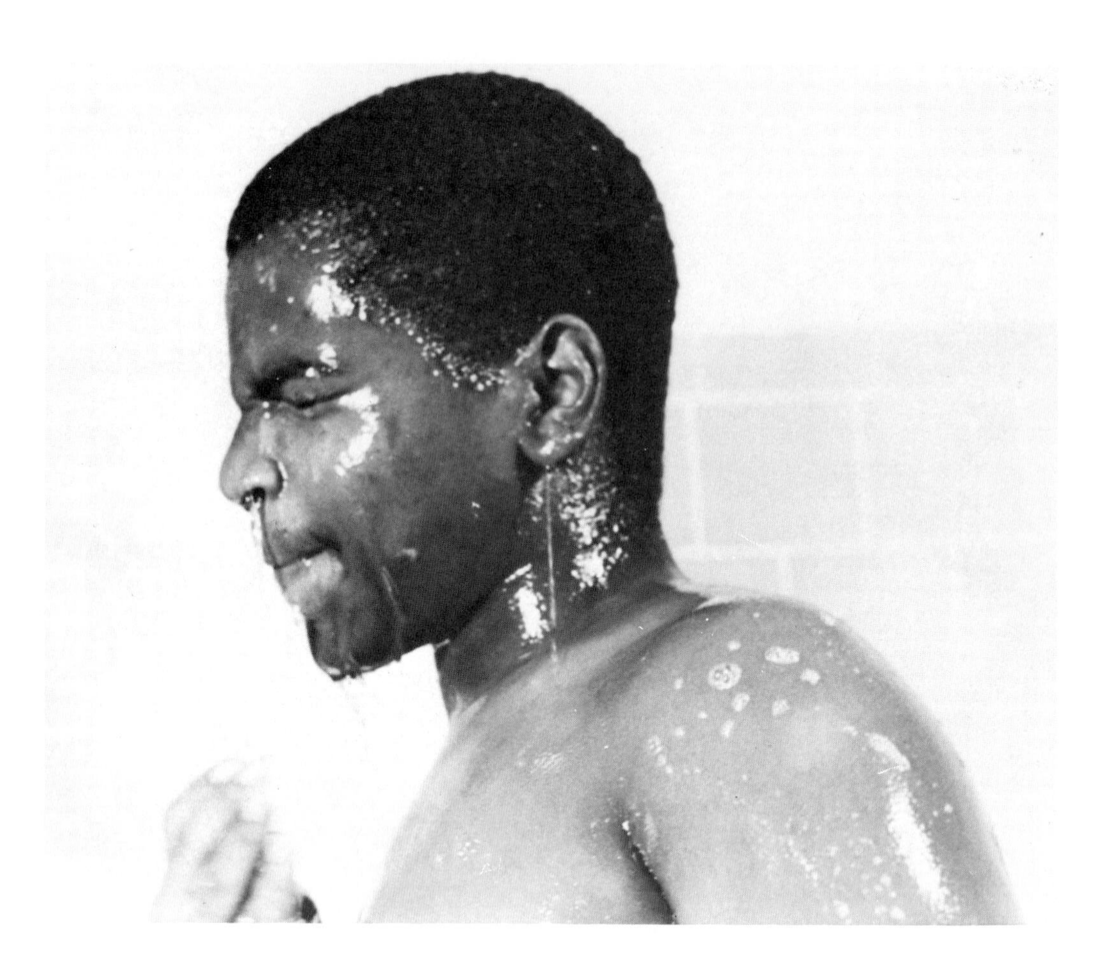

Water Everywhere

Look at a map of the world. Do you see more land or more water on the map?

Right! You see more water. Three-fourths of Earth is covered with water.

Earth's living things need that water. We must put water into our bodies every day. We drink it. Or we get it from the foods we eat.

People use huge amounts of water. We use it to raise our food. We use it to cook and to clean. We use it in our factories to make things such as clothes and cars. How else do we use water?

Water is a *renewable resource*. So is air. A renewable resource can be used again and again.

Even though Earth seems to have a lot of water, we can use only a small part of it. And we have problems with water that we can use. What do you think some of those problems are?

One problem is this: Some communities can't get enough water. Their water supplies are drying up.

Another problem is this: Water can become polluted. It can become polluted by chemicals and harmful germs. Why is water pollution dangerous?

Oceans in Motion

Here is something to think about: Most of the water you use comes from oceans. How do you think water travels from the ocean?

It travels in the air!

Suppose you fill a cup with water. You leave that cup of water in a safe place for a week. What happens to the water in the cup?

Right! It *evaporates*. It turns into water vapor and goes into the air.

Water on Earth is always evaporating. Huge amounts evaporate from the oceans. The water vapor rises. It changes into tiny drops of water. Those tiny drops form clouds.

The clouds move. As they move, they pick up more tiny drops of water. The clouds get bigger. What moves them?

Right! Winds move the clouds.

Winds blow the clouds over the land. Rain, ice, and snow fall from the clouds. They fall to the ground.

Some of that water goes into lakes. Some goes into rivers. And some soaks into the ground. Your community pipes that water to your house from one of those places.

That's how water from an ocean gets to your house!

Now look at the diagram. It shows how water moves from the ocean to the land.

Glade Walker, U.S. Bureau of Reclamation

A reservoir

San Jose Mercury News

A water tank

Our Sources of Water

Every community must have a water **source**. That's the place where the community gets its water.

What do you think some water sources are?

Lakes, rivers, and *ground water* are some water sources. (Ground water is water that collects under the ground.)

Small communities often get their water from underground sources. They dig wells. The wells reach the underground water and bring the water to the top of the ground.

Large communities usually get their water from lakes and rivers. If those communities aren't near a lake or a river, their water must be piped in from far away.

Communities must make sure they always have enough water. So many of them *store* water. That means they keep extra water.

One way to store water is to **dam** it. A dam is a huge, strong wall that stops a river from flowing. Water piles up behind the dam and forms a **reservoir**. A reservoir is a kind of lake.

Communities that have wells store water differently. They use big tanks like the one shown in the picture at the bottom of the page. That water is used only when the community needs extra water. When could that happen?

Let It Flow!

Water flows through many places before it reaches your home. The diagram on this page shows some of those places. Follow the diagram. Answer these questions about it. Then check your answers. (The right answers are upside down.)

1. Water flows from a kind of lake that a dam formed. What is that lake called?
2. Before people can use the water, it must be *treated*. So the water flows to a **treatment plant**. There water flows through *filters*. The filters clean out leaves, dirt, and other particles. Next, chemicals such as *chlorine* are put into the water. What do you think chlorine does?
3. The treated water then goes to a **pump station**. What do the pumps push the water through next?
4. Big pipes carry water from the pump station to your community. Smaller pipes then carry the water into **your house**. Those pipes are hooked up to things that use water. What are some of those things?

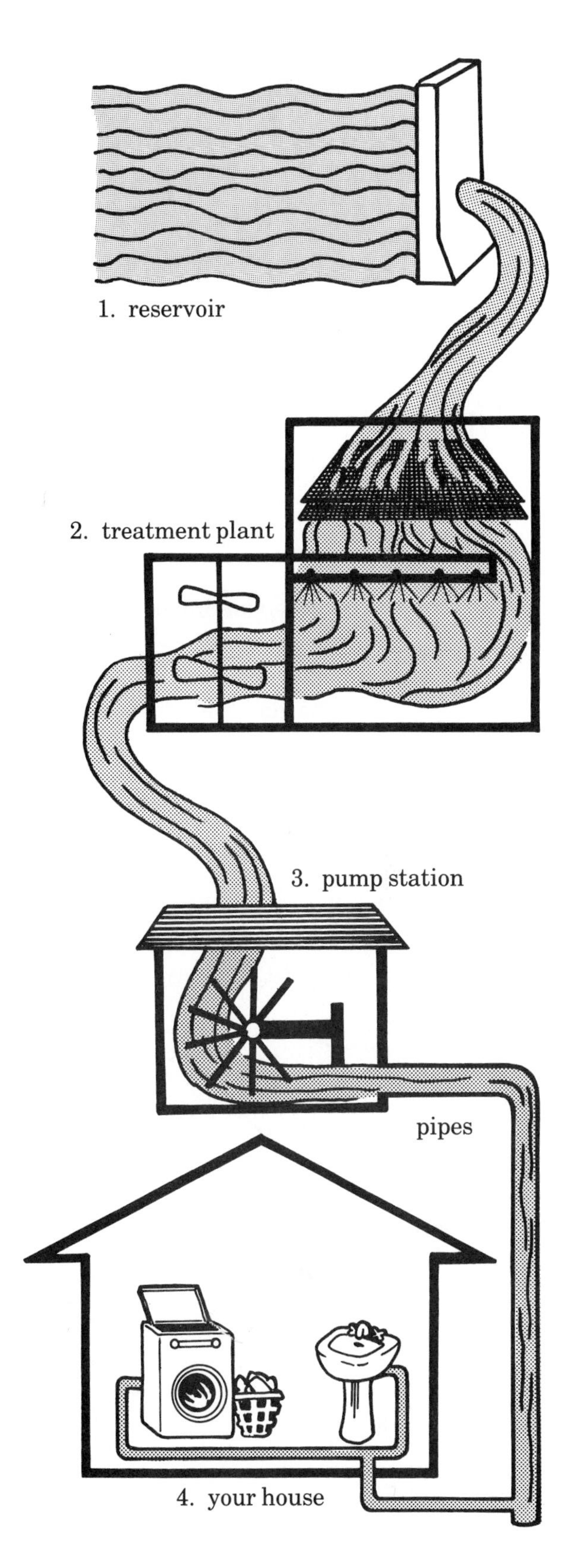

Answers

1. reservoir 2. kills the germs 3. pipes
4. sinks, showers, tubs, toilets, clothes washers, dishwashers

Why Is Water Treated?

You learned that the water you drink has been treated. Find out why. You'll need these things:

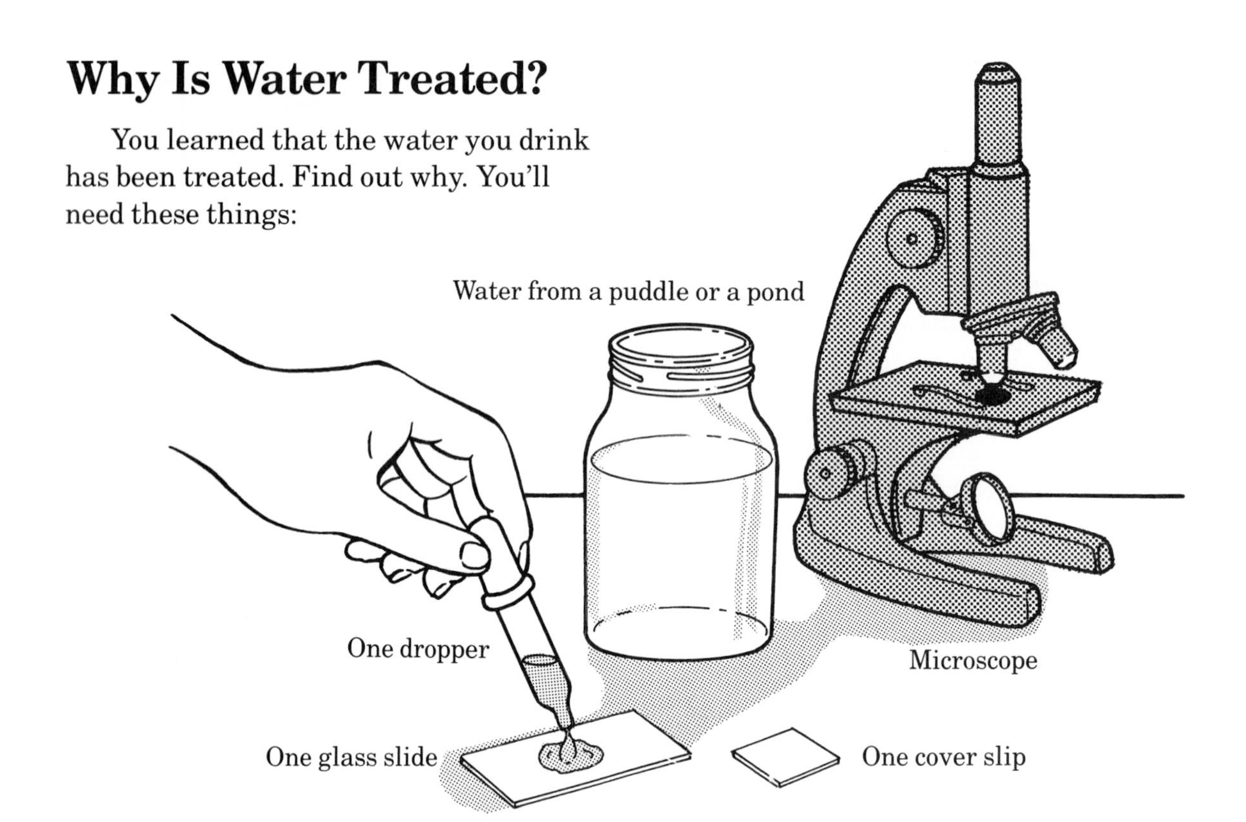

1 Put three drops of water on the slide.

---

2 Hold the cover slip at its edges. Hold it at one end of the slide, the way the picture shows.

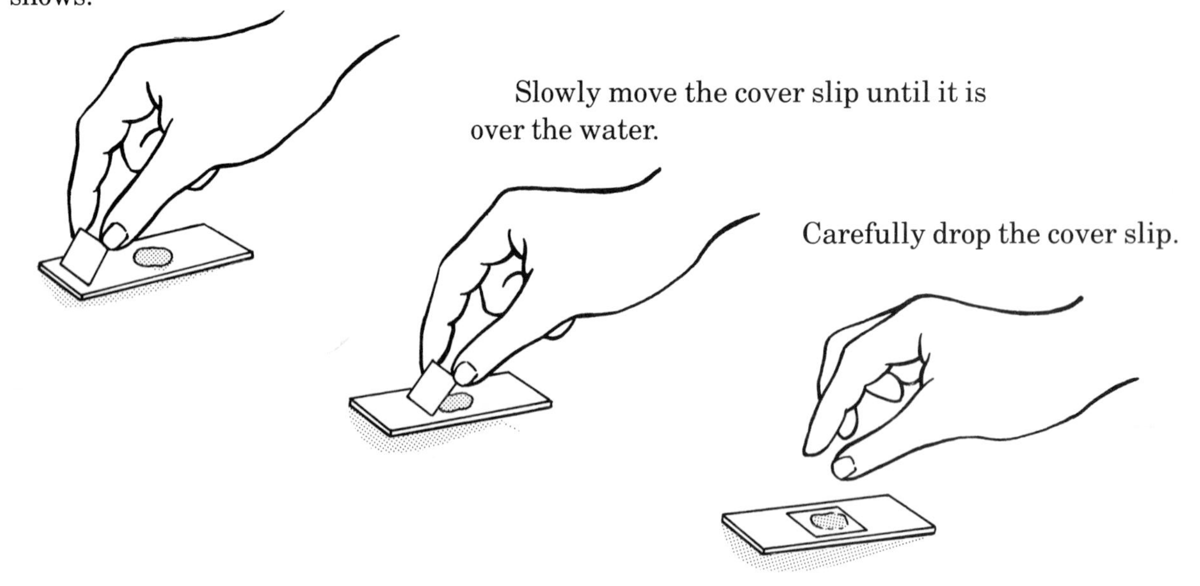

3 Put the slide in the microscope. Turn to low power. Look into the lens and *focus* it.

4 On a separate piece of paper, draw what you see.

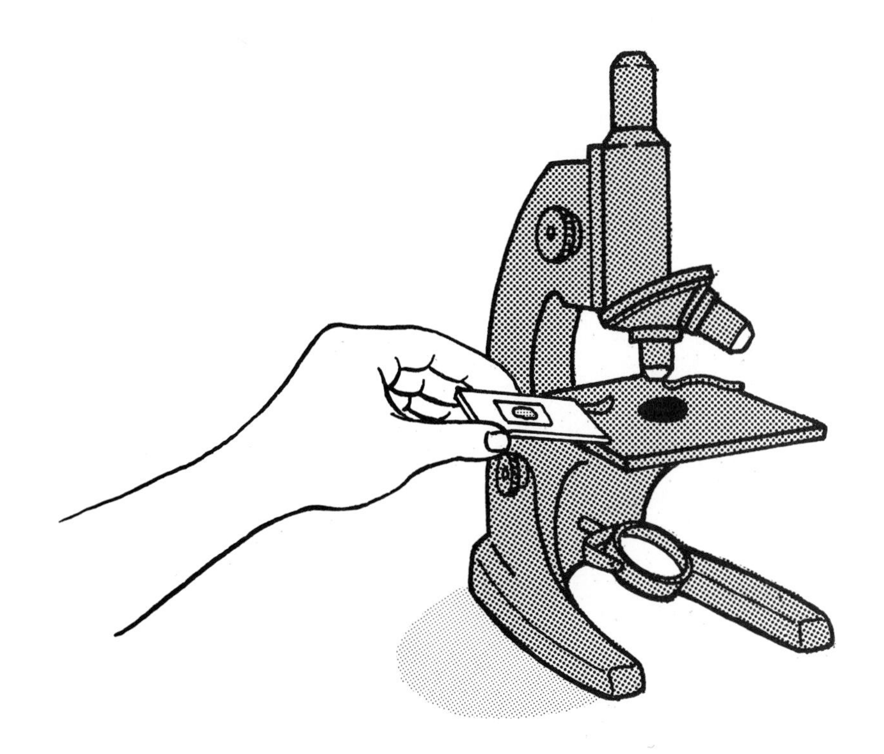

What Do You Think?

1. What do you see in the water?
2. What do you think those things are?
3. Why do you think we treat water?

Earth Watch

Your community government has a *water department.* That department makes sure your community's water is safe to use.

Look up the water department in a phone book. (Look under the name of your county or city.) On a separate piece of paper, write the name, address, and phone number of that department.

Talk with or write to someone in your water department. Ask these questions. Keep a record of what you find out.

1. How do people check the safety of our community's water?
2. What causes pollution in our community's water?

Review

Show what you learned in this unit. Choose the right words to finish each sentence.

1. Water is a vital resource because
   a. it moves around Earth.
   b. it is under the ground.
   c. it keeps living things alive.
2. Communities get water from sources such as
   a. wells, lakes, and rivers.
   b. farmland.
   c. rocks.
3. Water is treated to
   a. make more of it.
   b. make plants grow.
   c. make it clean and safe to use.

Check These Out

1. What water problems does your community have? (Examples are pollution, floods, and droughts.) Find an article in a local newspaper about a water problem. Read it and tell your class what you found out.
2. As you work through this section, you may want to find out more about Earth resources. You can find out more by looking in an encyclopedia or getting books from a library. You can also talk to an expert, such as a science teacher, an environmental worker, or a government official.

   Here are some things you may want to find out:
   - What is irrigation? How is it done?
   - What is desalinization of ocean water? How is it done?
   - Why do some people use biodegradable detergents?
   - How do lakes and rivers get polluted?

Unit 3

Living Off the Land

Air and water are vital resources. The land we live on is a vital resource too. It gives us all the things we need for food, shelter, and clothing.

Many of the things we use come from living things. We raise most of those living things. We raise them on large pieces of land.

- How do we use living things?
- Why is forest land important?
- Why is farmland important?

You'll learn the answers in this unit.

Before You Start

You'll be using the science words below. Find out what they mean. Look them up in the Glossary. On a separate piece of paper, write what the words mean.

1. **nutrient**
2. **topsoil**

U.S. Bureau of Land Management

Food, Clothing, and a Home

Your body needs air and water to stay alive. What else does your body need?

Right! Your body needs food. If living things can't get food, they will die.

We get all our food from plants and animals. Plants give us foods such as fruits, vegetables, and grains. Animals give us meat, milk, and eggs.

We use plants and animals in other important ways too. We use them as materials for making things we need. We use them to make clothing, shelter, and other things that we use.

Many of the materials we use come from plants. We use all parts of a plant: stems, leaves, seeds, fruits, and flowers. Here are some things we make from plant materials: cotton clothes, paper made from trees, wood furniture, wood houses, and medicines.

What other things do we make with plant materials?

Some animal materials we use are skins, furs, and feathers. Here are some things we make with animal materials: leather coats, leather shoes, feather pillows, and fur coats.

What other things do we make with animal materials?

Food

Can you name a food that doesn't come from a living thing? You probably can't, because there isn't any. Salt doesn't count.

Think about a pizza. The crust is made of flour. Flour comes from the seeds of the wheat plant. The sauce is made of tomatoes. Tomatoes are the fruit of tomato plants. The cheese is made from cow's milk.

Many different plants give us food. Each has a certain part that we eat. We eat the leaves of spinach and the roots of carrots. We eat the seeds of beans and grains and the fruits of apple trees.

We use many different animals for food. We eat chickens' eggs and cows' milk. And many people eat the flesh of fish, chickens, cows, pigs, and other animals.

On a separate sheet of paper, make a list of your favorite foods. For each, write down all the ingredients you know. Put the ingredients in two columns, *From Animals* and *From Plants*. Then for each ingredient, write down the kind of plant or animal it comes from. Add the animal or plant part if you know it.

Forest Land

Huge areas of Earth's land are thickly covered with trees and other plants. Those areas are called **forests**. (In some parts of the world, they are called *jungles*.) Forests are an important source of wood. But forests are also important to life. Why is that?

One reason is that forests are homes for many kinds of wild animals. Without the forests, some of those animals would become *extinct* (die out).

Forests are important in another vital way: Forests are made up of hundreds and hundreds of green plants. (Trees are one kind of green plant.) Green plants put gases that we need into our air.

One gas is oxygen. Plants take carbon dioxide from the air. They use it to make their food. While they make their food, the plants also make oxygen. The oxygen goes out of the plant and into the air.

Why is oxygen important to us?

Right! We breathe oxygen to stay alive.

Plants also put water vapor into the air. The roots of the plants take in water from the soil. The water moves to the leaves, where some of it is used. The extra water goes into the air as water vapor. That water vapor becomes clouds. The clouds turn into rain, ice, or snow.

Many forests all over the world are being cut down. People cut down trees for lumber. They clear the forests to build new towns. What do you think could happen if we cut down too many forests?

U.S. Bureau of Land Management

Farmland

Where do we get the plants and animals that we use for food or materials?

Right! We raise them. We raise *crops* (plants) and *livestock* (animals) on farms.

Farms are pieces of land that are used just to raise things. If we didn't have farmland, we wouldn't have food to feed the millions of people in our country. We wouldn't have enough materials to make things.

Farm crops can grow only on a certain kind of land. That land must be fairly flat. And it must have *topsoil*.

Topsoil is dirt that plants can root in easily. That dirt must also have the *nutrients* that plants need. And it must be able to hold a certain amount of water.

A farm can lose its topsoil. For example, the topsoil can get very dry during a **drought**. It can blow away. What do you think a drought is?

Right! A drought is a time when there is no rain for many, many months.

Suppose a farm has a lot of rain. The rain floods the land and turns it into a river. How can that be bad for the farmland?

Yes! The topsoil can be washed away.

What other problems can farmland have?

J.C. Dahilig, U.S. Bureau of Reclamation

Acid Rain

You've learned that plants grow in topsoil. Topsoil has the right nutrients. When it rains, water dissolves the nutrients. Plants take in the water and the dissolved nutrients.

Some people are becoming worried about our topsoil. They say it is becoming polluted. That pollution happens because two other Earth resources are also polluted. What do you think those resources are?

Right! The polluted resources are air and water. Sometimes pollution travels through the air or water for hundreds of miles.

When we burn things such as coal, pollution goes into the air. (We burn a lot of coal in our factories.) One form of that pollution is a certain gas. When the gas mixes with water, it forms an acid.

That happens when it rains. The gas mixes with rainwater and makes an acid. The acid falls down with the rain. It goes into the topsoil. That rain is called *acid rain*.

Acid rain is a serious pollution problem. It puts too much acid into the soil. That can harm plants.

What happens when there's too much acid in soil?

Experiment 2

What does too much acid do to plants?

Materials

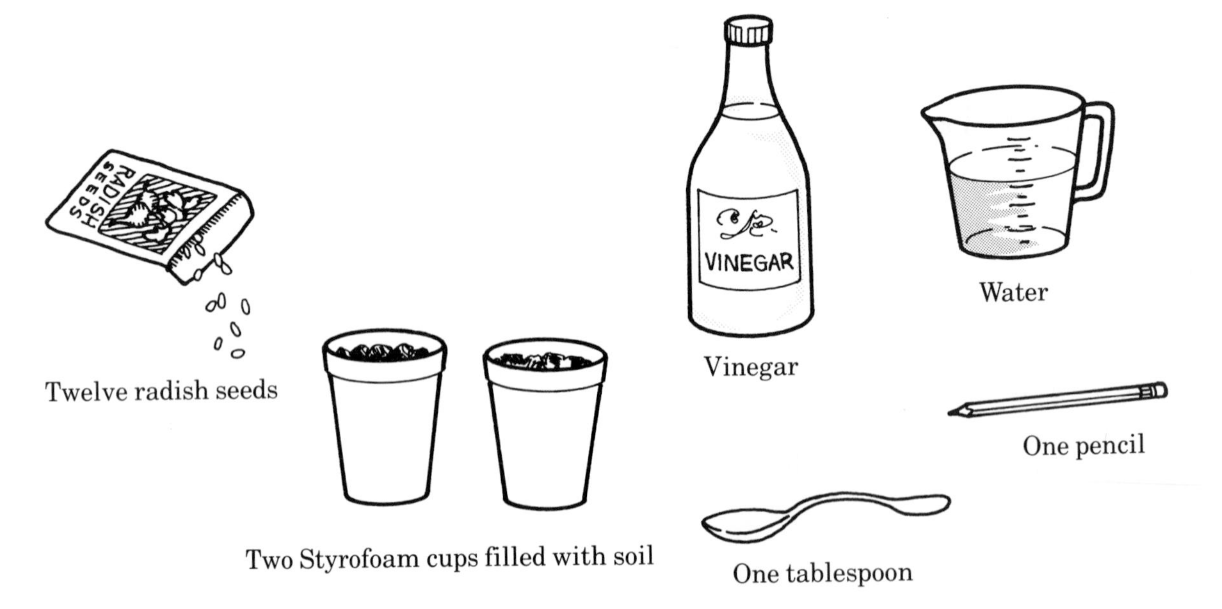

Twelve radish seeds

Two Styrofoam cups filled with soil

Vinegar

One tablespoon

Water

One pencil

Procedure

1. Write *acid* on one cup. Write *no acid* on the other. Plant six seeds in each cup.

2. Pour a little water into both cups.

3. Pour two tablespoons of vinegar in the cup marked *acid*.

4. Put the cups in a safe place. Wait for three days. Then look at the cups.

Observations

1. After three days, what happens in the cup that has acid in it?
2. What happens in the cup that doesn't have acid in it?

Conclusions

How does acid affect plants?

Earth Watch

Many communities have government *agriculture* agencies. Those are offices that watch over farming. Two such agencies are the U.S. Department of Agriculture and the State Agriculture Commission.

Look up an agriculture agency in a phone book. On a separate piece of paper, write the name, address, and phone number of that agency.

Write to or talk with someone at the agency. Ask these questions. Keep a record of what you find out.

1. What crops grow in our community?
2. Which fertilizers, pesticides, and herbicides do the farmers use?
3. Where do they get water for crops?
4. What is done to protect the topsoil?

Review

Answer the questions below. Check your answers by looking at the page listed after each question.

1. How do we use plants? (page 160)
2. How do we use animals? (page 160)
3. Why are forests important? (page 162)
4. Why is farmland important? (page 163)
5. How is topsoil becoming polluted? (page 164)

Check These Out

1. Bring some soil to class. Plant some seeds in it. Find out if it is good for growing things.
2. Visit a park or wilderness area. Draw some plants and animals you see. Find out their names.
3. Here are more things you may want to find out:
   - What are some ways that farmers get plants and animals to produce more food?
   - What are some problems that ecologists study?

Unit 4

Fossil Fuels

You learned about air, water, and land. Those are vital resources. We need them to stay alive. Now you'll be learning about another resource—fossil fuels. Fossil fuels are not vital resources. But they are very important to us. Fossil fuels give us the energy to run our machines. They light our buildings, heat our houses, and cook our food. They make life easier and more comfortable.

- What are fossil fuels?
- How do we get fossil fuels?
- How do we use fossil fuels?

You'll learn the answers in this unit.

Before You Start

You'll be using the science words below. Find out what they mean. Look them up in the Glossary. On a separate piece of paper, write what the words mean.

1. **conserve**
2. **fossil**
3. **properties**

Ancient Energy

Every time we turn on a light, we use energy. Every time we use an appliance or run a machine, we use energy. We get most of that energy by burning fossil fuels. Coal, petroleum (oil), and natural gas are fossil fuels. Why do you think those fuels are called fossil fuels?

Those fuels are made from the *remains* (fossils) of plants and animals that lived millions of years ago.

Scientists think coal was formed this way: Millions of years ago, Earth was a much warmer place. It was covered with muddy swamps. Many large plants grew in those swamps.

As the plants died, they fell to the bottom of the swamps. Layers of dead plants built up. Then layers of soil and rock covered them. As years went by, more layers of soil and rock pressed down on them. Hundreds of millions of years passed. The plant material began to change. It became coal.

Coal gives off a lot of heat when it burns. It also gives off a lot of black smoke. It burns slowly, so it burns for a long time. We use it to heat buildings, to run trains, and to run our factories. We use it to make metals and electricity.

Like all fossil fuels, coal is found under the ground. It must be dug out. Mining companies dig mines to get the coal. They dig long tunnels under the ground. Or they dig up huge areas of land.

Coal is a resource that can be used up. So we say it is a **nonrenewable resource**. When we take it out of the ground, no new coal forms to take its place.

What problems do you think we have with coal?

Chevron U.S.A. Inc.

Chevron U.S.A. Inc.

Petroleum and natural gas are found all over Earth. People dig under land and under oceans to get them.

Ocean Fossils

Petroleum and natural gas are also fossil fuels. Scientists believe those fossil fuels started forming about 700 million years ago. Here is how: Many tiny plants and animals lived in the oceans. When they died, their remains sank to the ocean floor. Slowly, over millions of years, they were pressed into the earth. They changed into petroleum and natural gas. Some of the oceans disappeared. They became dry ground.

Petroleum and natural gas are found deep under the ground. How do you think we get them to the surface of the earth?

We get petroleum and natural gas by digging wells. A long hole is drilled through the ground. It reaches the petroleum and natural gas. Then a powerful pump forces them up to the top of the ground.

We use natural gas for cooking and for heating buildings. We use petroleum for many things. It comes out of the ground as a thick black liquid. We call that liquid *crude oil.* Petroleum is made into products such as gasoline, kerosene, and butane. What are some other petroleum products?

Plastic, paint, nylon, and polyester are all petroleum products. Many roads are made from a petroleum material called *asphalt.* And all cars use motor oil, another petroleum product.

Experiment 3

What are the properties of oil?

Petroleum has certain properties. Those properties make petroleum good to use in machines. Machine oil is a product of petroleum. It has some of the same properties as petroleum. What are those properties? Find out!

Materials

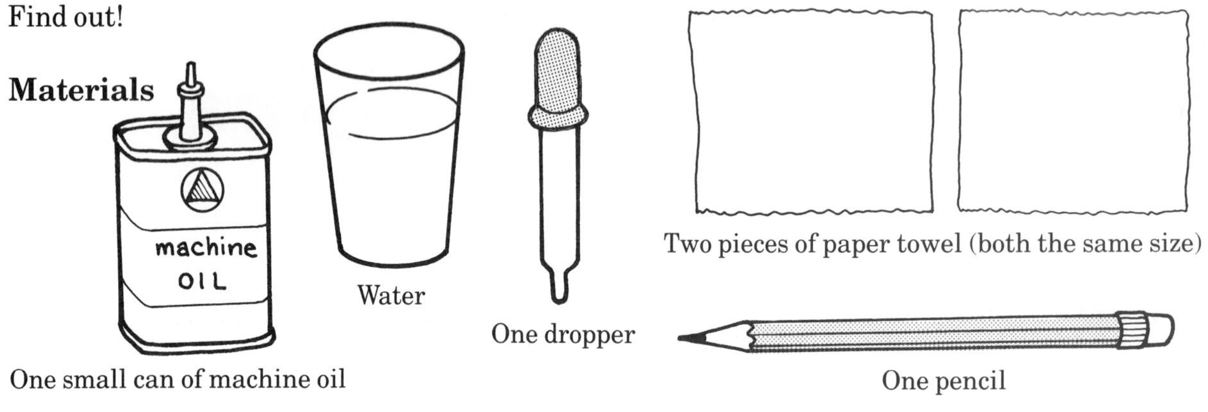

One small can of machine oil

Water

One dropper

Two pieces of paper towel (both the same size)

One pencil

Procedure

1. Write *oil* on one piece of paper towel. Squeeze a drop of oil on it.

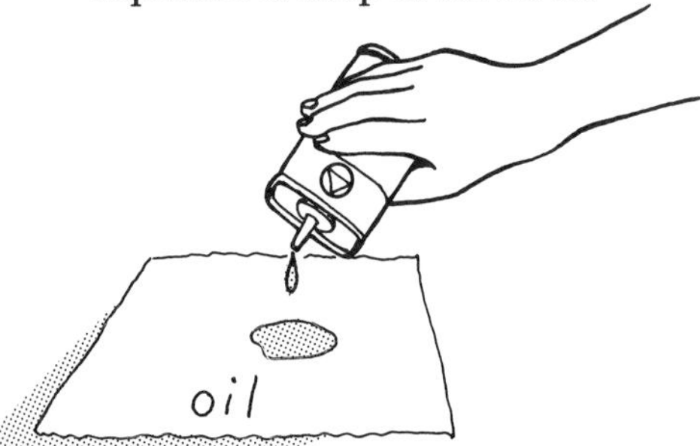

Write *water* on the other piece of paper towel. Put a drop of water on that paper towel.

2. Put the two paper towels under a strong light or by a sunny window. Leave them there to dry.

3. Now squeeze a few drops of oil on your finger. Rub your thumb and finger together. How does the oil feel?

4. Rinse your fingers in the glass of water. Does the oil come off?

Observations

1. How does the oil feel on your fingers?
   a. It feels dry.
   b. It feels slippery.
   c. It feels hot.
2. What happens to the oil when you rinse your fingers?
   a. It stays on your fingers.
   b. It washes off.
   c. It turns brown.
3. Look at the paper towels. Which did *not* evaporate: oil or water?

Conclusions

Machine oil has these properties: It is slippery. It takes a long time to dry out. And it doesn't wash off. So it makes things move more easily. And it stays on things for a long time. How do you think oil is used in machines?

The Trouble with Fossil Fuels

Petroleum and natural gas are nonrenewable resources. Why do you think they are nonrenewable?

They are nonrenewable because they take millions of years to form.

We have already used up much of the petroleum and natural gas in our wells. So oil companies must now dig deeper for those resources. They must also find new places to dig wells. Some people think that someday we'll run out of places where we can easily get those fossil fuels. What do you think will happen then?

Here are some other problems. We damage other resources when we get fossil fuels out of the ground. For example, coal mining takes away topsoil and leaves big holes. Petroleum is moved on ships or through pipes. It can accidentally spill into places such as an ocean. When that happens, it kills plants and animals.

To use fossil fuels, we must burn them. That causes another big problem. What is that problem?

Right! That problem is air pollution and smog.

Smog moves with the air. It moves to our farmland. It moves to our lakes. Chemicals from smog get into our crops, animals, lakes, and fish.

Those farm crops, farm animals, and fish are often our food. Many of them now have dangerous chemicals in them because of pollution. What do you think we should do to make our food safer?

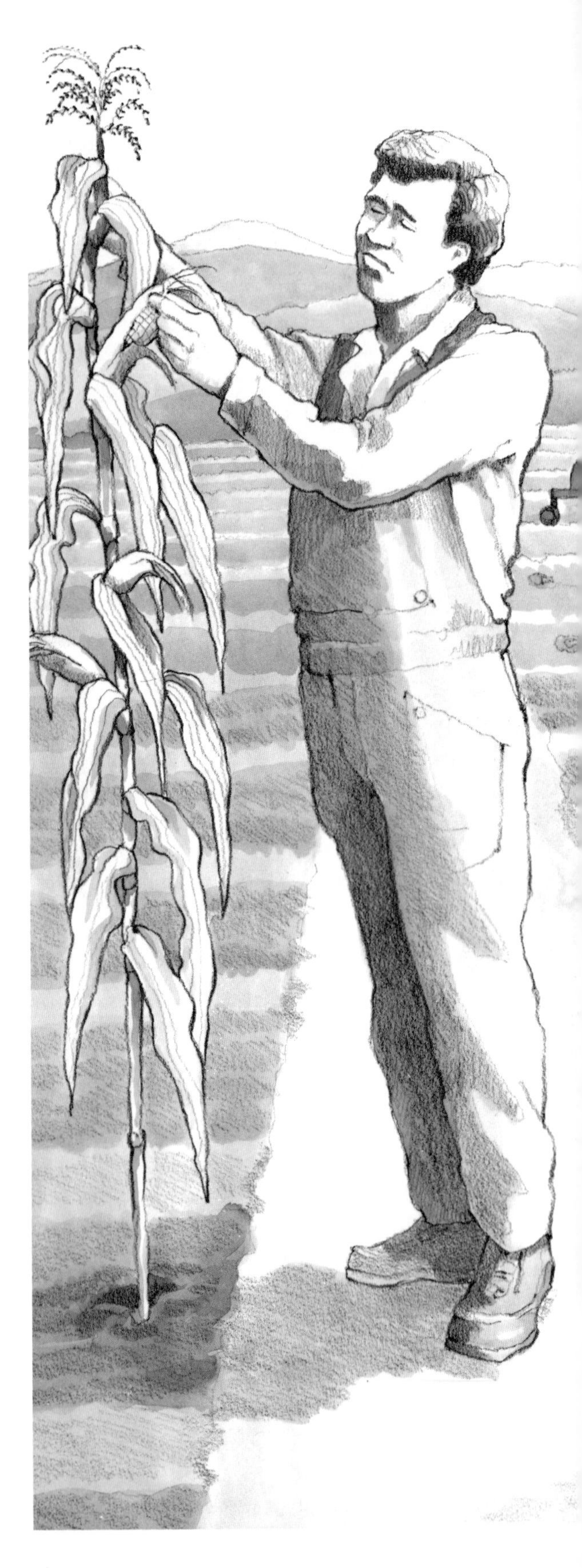

What Can We Do?

Polluted air is a serious problem. So many businesses are trying to clean up our air. How do you think they are doing that?

Here are some ways:

Most factories put special equipment on their smokestacks. That equipment traps harmful chemicals and keeps them from going into the air.

The companies that make cars do this: They put *pollution control devices* on cars. With pollution control devices, cars make fewer harmful chemicals.

Sometime in the future, we may have another serious problem. We may run out of fossil fuels. So people are beginning to *conserve* fuel. They use less of it. People conserve fuel at home, and they conserve fuel on the road. What are some ways that you can conserve fuel?

Right! Turn off lights, televisions, and stereos that aren't being used. Keep the refrigerator door closed. Put on extra clothes when you are cold, instead of turning up the heat.

When you travel, you can conserve fuels these ways: Drive with other people, or take a bus. Whenever possible, ride a bicycle or walk.

Earth Watch

Communities have *utility companies*. Those companies supply fuel—heating oil, natural gas, and coal—to the homes and businesses in a community.

Find out where a utility company is in your community. Look up *gas company* or *power company* in the yellow pages of your phone book. On a separate piece of paper, write the name, address, and phone number of the company.

Write to or talk with someone at that company. Ask these questions. Keep a record of what you find out.

1. How is coal used in our community?
2. How is natural gas used in our community?
3. Which petroleum fuels are used in our community? How are they used?

Review

1. Each sentence below tells about a fossil fuel. Which fossil fuel is it: coal, petroleum, or natural gas?
   a. Gasoline, machine oil, and many other products are made from this fossil fuel.
   b. This fuel is used for cooking and heating.
   c. This fuel is burned to make steel and electricity.
2. Why are fossil fuels called nonrenewable resources?

Check These Out

1. What did the world look like during the time of dinosaurs? Draw a picture.
2. Make a list of things that use fossil fuels. How many things did you list?
3. Here are more things you may want to find out:
   - What are petrochemicals? How are they made? What are they used for?
   - Why is coal mining a dangerous job? How do coal miners protect themselves from danger?

Unit 5

Minerals

Have you ever watched a big building go up? First a concrete floor is poured. Then steel beams are put in. Iron bolts fasten the beams together. And more concrete is poured for walls.

Concrete floors, steel beams, and iron bolts are made from *minerals*. Minerals are a valuable resource. We use minerals to make buildings and many kinds of products. Our way of life depends on those products.

- What kinds of minerals do we use?
- What materials and products do we make from minerals?
- What problems are caused by our use of minerals?

You'll learn the answers in this unit.

Before You Start

You'll be using the science words below. Find out what they mean. Look them up in the Glossary. On a separate piece of paper, write what the words mean.

1. **crystal**
2. **raw materials**
3. **recycle**

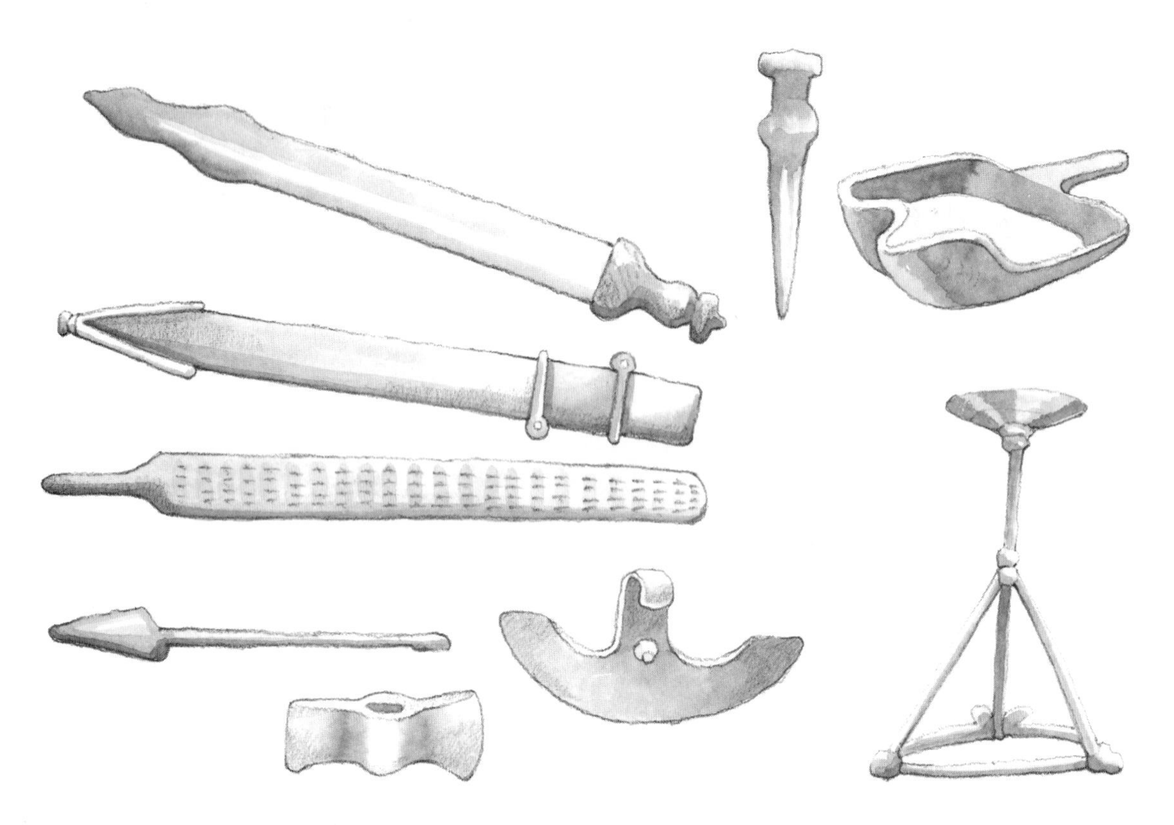

Raw Materials

Thousands of years ago, people dug a mineral from the ground. It was their raw material for making spears and knives. That mineral was iron. Today we still use minerals as raw materials. We use them in almost everything we make.

We say minerals are *metallic* or *nonmetallic*. Metallic minerals have **metals** in them. They are usually heavy. Iron, silver, tin, aluminum, copper, and gold are examples of metallic minerals. We melt the metals, then form them into products. What's a product that is made of metal?

Nonmetallic minerals are not as heavy as metallic minerals. Many are very light. Some look like glass. Some join together to form rocks. Quartz, salt, gypsum, and marble are examples of nonmetallic minerals. So are gems, such as diamonds and rubies.

We use quartz to make things such as microscope lenses. We put salt in our food. We make plaster with gypsum. And we use marble for things such as sinks, walls, and statues. How do we use diamonds?

Certain nonmetallic minerals can be crushed and mixed together to form very strong materials called cement and concrete. How do we use those materials?

Minerals are a nonrenewable resource. We can run out of them. All our mineral resources are found in the ground. We must mine them to get them out.

What problems do you think we have with minerals?

Crystals

Earth minerals are made of crystals. Crystals start as very tiny shapes. If more mineral material is added, crystals "grow"—get bigger.

See how the crystals of one mineral grow. You'll need these materials:

- Salt
- One dish
- ½ cup hot water
- One spoon
- One hand lens

1 Pour a little salt into the water. Stir until the salt disappears. Add more salt. Stir and add salt until it no longer disappears.

2 Let the mixture cool. Then pour some on the dish. Put it in a safe place, and wait for two days.

3 After two days, look at the dish. You should see salt crystals. Look at them with the hand lens. On a separate piece of paper, draw one of the crystals.

San Jose Mercury News

Cars are made of many different metals. What do you think some of those metals are?

Metals in Your Life

Think of a gold ring. Think of aluminum foil. Both are products that are made of metal. When we think of a metal, we think of something that is hard, bright, and shiny. But metals don't always look like that when we dig them out of the ground. That's because they are mixed in rocks. We call those rocks **ores**.

Mining companies dig mines to get the ores. Next, factories take the metal out of the ores. That metal is then used to make things.

Metal is a very good material for making many different kinds of things. Why do you think that's so?

Metal can be melted or pounded. It can be formed into many shapes. And it is hard and strong. Things made from metal last a long time.

We use many different metals. One of them is copper. We make wire from copper. What else do we use copper for?

Aluminum is a good metal for many products. Cooking pots can be made from aluminum. What else is made of that metal?

Sometimes we mix metal and other minerals together. That makes a metal material called an **alloy**. Steel is an alloy. It is stronger than any one metal alone. We use steel to make many things. Paper clips, knives, and ships are some examples of steel products. What are some other things made of steel?

An Iron Factory

The metal we use the most is iron. But the iron in Earth is not in a form we can use. It's mixed with other minerals in ore. How do you think we take iron out of the ore?

Right! We melt the ore in an iron factory.

Here is how we get iron: First, iron ore is dug out of a mine. It is put in a machine that crushes it into small pieces. Then it is taken to an iron factory.

The iron ore is dumped into a huge *blast furnace*. That is a special kind of furnace that gets very hot. Another mineral and a special kind of coal are also added. They help the ore to melt and the iron to separate from it.

Iron is the heaviest mineral in the melted ore. It sinks to the bottom. And it is drained out an opening in the bottom of the furnace. Most of that iron will be used to make steel. So it is sent to another furnace where steel is made.

Many Earth resources go into making iron and steel. One resource is minerals. Two other resources heat the furnace. What do you think they are?

Another resource keeps the furnace from getting too hot. It also cools the steel. What resource do you think that is?

Right! Fossil fuels and air heat the furnace. Water cools the furnace and the steel.

San Jose Mercury News

Inside a steel factory

The steel in these wrecked cars will be recycled. It will be used again to make new steel.

San Jose Mercury News

Trouble Again!

We have problems with minerals! Some are like the problems we have with fossil fuels. What is one of those problems?

Every year, we want more and more mineral products. So more and more minerals are taken out of the ground.

Taking out minerals harms Earth. It leaves giant holes. Then no one can live where the holes are. No crops can grow there either. And land around the holes can cave in, making the holes even bigger!

Another problem with minerals is this: We could run out! Many minerals are getting harder and harder to find.

Many people are trying to save our mineral resources. How do you think they're doing that?

People *recycle* mineral products. That means they use the minerals in those products again. For example, old copper wires and pipes are melted. They are used to make new copper products. People also recycle steel products, such as wrecked cars. They are melted down, and the steel is used to make new cars.

You use many products that are made from minerals. You could recycle some of them. What is one product you could recycle?

Earth Watch

Most communities have a *recycling center.* That's a place where people bring throw-away products such as cans and bottles. Those products will be used to make new products.

Find out where a recycling center is in your community. Look under *recycling centers* in the yellow pages of the phone book. On a separate piece of paper, write the name, address, and phone number of the center.

Talk with or write to someone at that center. Ask these questions. Keep a record of what you find out.

1. What materials do you collect?
2. What do you do with the materials you collect?
3. What new products will be made from those materials?

Review

Answer the questions below. Check your answers by looking at the page listed after each question.

1. What are some metallic minerals? (page 176)
2. What are some metal products? (page 178)
3. What are some nonmetallic minerals? (page 176)
4. What are some things we make with nonmetallic minerals? (page 176)
5. What is one problem with minerals? (page 180)

Check These Out

1. Start a recycling project in your classroom. Bring in materials such as cans and bottles. Take them to the recycling center. Or arrange for the center to pick them up.
2. Make a rock collection. Pick up rocks in your neighborhood or around your school. Wash them and put them in a box. Talk to an expert and find out what kinds you have.
3. Find out what factories are in your community. What jobs are done in those factories?
4. Here are more things you may want to find out:
   - How do scientists identify minerals? How are minerals tested for hardness?
   - Some companies want to mine the ocean floor. What minerals would they get?

Unit 6

Save Your Resources!

Problems!

You learned about some of Earth's vital and important resources.

What vital resources keep us alive?

What resources give us energy?

What resources give us raw materials?

You also learned that we have problems with our Earth resources. What problems do we have with these resources?

- Air
- Water
- Forests and forest land
- Farmland
- Fossil fuels
- Minerals

Get the Facts!

Suppose your state, city, or town has a resource problem. Maybe it's air, water, or soil pollution. Maybe fuel is too expensive and too hard to get. Or maybe land is being washed away. That problem affects you!

How would you find out all you can about that problem?

Here are some good ways to find out about a resource problem in your community:

1. Look at newspapers and news magazines. What newspapers are in your community? What magazines are in your community?
2. Watch local TV news programs about your community. What TV programs can you watch?
3. Talk to people in environmental groups. Those are people interested in saving natural resources such as wild animals, forest lands, seashores, air, water, and so on. What environmental groups are in your community?
4. Talk to people in government agencies. (You already know of some agencies.) What agencies are close to your community?
5. Talk to people in businesses such as oil companies or power companies. What are some businesses like those in your community?

What's the Problem?

Choose one resource problem that you'd like to know more about. Then get the facts about that problem. Keep a record of what you find out. On a separate piece of paper, answer the questions below. Make a list of all the things you read and saw. List all the people you spoke with or wrote to.

1. What is the problem?
2. What causes the problem?
3. What resources are harmed by the problem? How are those resources harmed?
4. How are the people in your community harmed by the problem?

Take Action!

Now you know about a problem your community has. You may want to do something about it. You probably won't be able to solve the problem. But at least you can make it better. On a separate piece of paper, answer these questions.

1. What is the problem?
2. What do you think can solve this problem?
3. What can the people of your community do about the problem?
4. Who can help the people of your community with this problem? (Think of businesses, government agencies, and environmental groups.)

Show What You Learned

What's the Answer?

1. Which resource is it? Read each clue, and then choose the right resource or resources from the list below. There may be more than one right answer.

   air — fossil fuels
   minerals — plants and animals
   water

   a. We use this resource for drinking, washing, and growing crops.
   b. This resource is made of gases.
   c. We make things from these resources.
   d. These resources are nonrenewable.
   e. These resources keep us alive.
2. Choose one resource. Explain what problems we can have with it.

What's the Word?

Give the correct word or words for each meaning.

1. The air that is around Earth
   A ________
2. Harmful material that gets into our resources
   P ________
3. To change from a liquid to a gas
   E ________
4. To use less of a resource
   C ________
5. A mixture of metals, such as steel
   A ________
6. What we use to make things
   R ________ M ________
7. Layer of soil that plants grow in
   T ________

Congratulations!

You've learned a lot about Earth resources. You've learned

- What resources help keep us alive
- What resources we make things with
- How your community uses resources
- How people protect and conserve our resources
- And many other important facts about the planet we live on

Glossary
The Solar System

as ter oids Pieces of rocky material that travel in space and revolve around the sun

as tro naut A person who travels in space

at mo sphere Layer of gases around a planet

ax is The straight line around which something spins

col lide To crash with

com ets Frozen balls of ice, gases, rock, and dust that have long orbits around the sun

cra ter A hole in the ground that's shaped like a bowl

el lipse A shape like an oval

gal ax y Billions of stars that form a group in space

grav i ty A strong pull that all planets and stars have

me te or ites Rocks and dust from space that have crashed into planets

me te or oids Rocks and dust that travel through our solar system

me te ors Rocks and dust from space that burn up in Earth's atmosphere

NASA National Aeronautics and Space Administration; the United States program to explore space

ob ject Thing

or bit The path a thing makes as it moves around another thing

plan ets Worlds such as Earth and Mars

re flect To bounce off something

re volve To move completely around something

ro bot space craft A spaceship without people that travels through space

ro tate To spin around

sat el lite A small thing that moves around a larger thing

sea sons The four parts of the year—winter, spring, summer, and fall

Sol One of the names for the sun

so lar sys tem The sun and all the things that move around it; our star system

space probes Trips through the solar system to gather information about space

space craft A spaceship

star A huge ball of superhot gases that gives off heat and light; a sun

star sys tem A star and all the things that move around it

tilt Tip or slant

Glossary

Changing Earth

at mo sphere The air that surrounds Earth

col lide To crash or run into something

con ti nent A very large piece of land that's mostly surrounded by water

core The center of Earth

core sam ple A sample of the Earth from the bottom of a lake or ocean

crust The rocky outside part of Earth

de pos it To drop out of water

dis solve To break down into very small pieces in water

e rode To wear away

e rupt To explode and push out of Earth

ev i dence Clues left by something

ex pand To take up more space

fault A deep crack on Earth

fos sil The remains of a plant or animal that lived a long time ago

fos sil fu els Fuels that are made from plants or animals that lived millions of years ago; coal, oil, and natural gas

ge ol o gist A scientist who studies Earth

gla cier A huge body of ice that moves across the land

green house ef fect The way the atmosphere traps the sun's heat

hu mid Having moisture in the air

ice age A time when much of Earth is covered with ice

ig ne ous rock Rock that forms from melted rock

land form The shape of a piece of land

lay er A thickness of rock or dirt under or on top of another

mag ma Very hot melted rock that is inside Earth

man tle The part of Earth that is under the crust

met a mor phic rock Rock that is changed by heat and pressure deep inside Earth

mo raine A high hill that forms when a glacier melts

moun tain A tall, steep kind of land

plain A low, flat kind of land

pla teau A kind of land that is not tall or flat

plates Pieces of the outside part of Earth

pres sure A force or push

sed i ment Bits of rock

sed i men ta ry rock Rock that forms from bits of rock

seis mo graph A machine that records and measures earthquakes

sur face The top or the outside of something

val ley A long, low kind of land

vent An opening in the surface of Earth

Glossary
Weather

air mass Moving air that is hundreds of miles wide

air pres sure How hard air pushes down on anything it touches

an e mom e ter Something that measures the speed of winds

ba rom e ter Something that measures how heavy air is

block A large amount

cloud A large group of tiny water drops that float together in the sky

con dense To change from a gas to a liquid

de grees Amounts of heat in the air

e vap o rate To change from a liquid to a gas

fog A large group of tiny water drops that float together near the ground

fore cast A report of what the weather will be

fro zen Turned into ice

hail Balls of ice that fall from the sky

hu mid Having a lot of water vapor in the air

hu mid i ty The amount of water vapor in the air

hy grom e ter Something that measures the amount of water vapor in the air

in stru ment Something that measures

mea sure To find out the amount of something

me te o rol o gist A scientist who studies the weather

miles per hour The speed of wind

pre cip i ta tion Water that falls from the sky

pre dict To guess what might happen

rain Drops of liquid water that fall from the sky

ro tate To spin

sleet Ice drops that fall from the sky

snow Frozen water vapor that falls from the sky

tem per a ture How hot or cold something is

ther mom e ter Something that shows how much heat is in the air

tor na do A very strong, spinning windstorm

wa ter va por The gas form of water

wind Air that moves quickly

Glossary
Earth Resources

al loy A metal that is made by mixing metals and other materials; steel is an alloy

at mo sphere The air that surrounds Earth

car bon di ox ide A gas in Earth's air. *Plants need carbon dioxide.*

con serve To use less of a resource so it lasts longer; to use a resource wisely

crys tal A certain shape that minerals form

dam To hold back water with a wall

drought A long time during which no rain falls on a place

e vap o rate To change from a liquid to a gas

for est A large piece of land that is covered with trees and other plants

fos sil The remains of a plant or animal that lived long ago

ground wa ter Water that collects under the ground

met al A hard shiny mineral that's used to make things

non re new a ble re source A resource that can be used up

nu tri ent Something plants take in to stay healthy; a food

ore Rock that contains minerals

ox y gen A gas that animals need. *Oxygen is one of the gases in air.*

par ti cle A very tiny piece of something

pol lu tion Harmful material that goes into air, water, and land

prod uct Something that is made from materials

prop er ties The ways that something looks or behaves

raw ma te ri als Things that we use to make other things

re cy cle To use something over again

re new a ble re source A resource that can be used again and again

res er voir A kind of lake where people store water for later use

smog Waste material from burning things that goes into the air

source A large supply of something that we use, such as water

top soil The layer of soil that plants grow in

vi tal Necessary for life

wa ter va por Water that's in the form of a gas